The IUCN Forest Conservation Programme

Conserving Biological Diversity in Managed Tropical Forests

Edited by

**Jill M. Blockhus, Mark R. Dillenbeck,
Jeffrey A. Sayer and Per Wegge**

Proceedings of a Workshop held at
the IUCN General Assembly
Perth, Australia
30 November – 1 December 1990

IUCN / ITTO
1992

Published by: IUCN, Gland, Switzerland, and Cambridge, UK
                 in collaboration with ITTO.

                 IUCN and ITTO are grateful to the Governments of the United Kingdom,
                 The Netherlands and Japan, who contributed to the cost of this study.

ITTO

Citation: IUCN (1992). *Conserving Biological Diversity in Managed Tropical Forests*. IUCN,
                 Gland, Switzerland and Cambridge, UK. xi + 244 pp.

ISBN: 2-8317-0101-5 ✓

Printed by: Unwin Brothers Ltd, Woking, Surrey, UK

Cover design by: James Butler

Cover photo: Timber elephants in the Pegu Yomas National Park in Burma/Myanmar: Jeffrey A. Sayer.

Produced by: The IUCN Publications Services Unit, Cambridge, UK.

Available from: IUCN Publications Services Unit
                 181a Huntingdon Road, Cambridge CB3 0DJ, UK
                 or
                 IUCN Communications Division
                 Rue Mauverney 28, 1196 Gland, Switzerland

The text of this book is printed on Reprise 80 gsm recycled paper.

TABLE OF CONTENTS

PREFACE

At its 8th session in Bali, Indonesia, in 1990, the International Tropical Timber Organization (ITTO) adopted the target of ensuring that all tropical timber marketed internationally should, by the year 2000, come from forests that are managed sustainably. It is now generally accepted that the term "sustainable" must apply not just to the maintenance of timber yields, but to the maintenance of all the goods and services provided by the forests. One of the most important services to be maintained in tropical forests is the provision of habitat for the exceedingly rich fauna and flora of the biome. This biological diversity conservation function is primarily served by national parks and equivalent totally protected areas. However, such areas only cover a small proportion of the biome and they alone are insufficient to conserve the full range of diversity of the forests. A recent study by The International Union for the Conservation of Nature and Natural Resources/The World Conservation Union (IUCN) (Whitmore and Sayer in press) shows that deforestation, forest fragmentation and degradation are likely to cause the extinction of a large portion of the biodiversity of tropical moist forests in the future. The study also shows that the risks of extinction could be diminished if extensive areas of near-natural forest were to be managed for sustainably to produce timber and other forest products. However, this would only happen if the forests were managed in an ecologically sensitive way.

The present study is an attempt to determine whether the member countries of the ITTO have a legal and administrative basis for managing their production forests in ways which will allow these forests to contribute to biological diversity conservation. The study also attempts to assess the extent to which such biodiversity-friendly management is already applied on the ground. For each member country, studies were commissioned on the situation of production forests, their spatial relation to the totally protected area system and the extent of knowledge of the impacts of present management on biological diversity. These country studies were presented at a workshop held during the IUCN General Assembly, in Perth Australia, in December 1990.

The workshop was attended by about 200 tropical forest specialists from all over the world. The country papers were presented and discussed in three sub-sessions covering the three main tropical regions. Subsequently regional reviews based on the conclusions of these sessions were presented to all workshop participants. These reviews are reproduced here along with country studies for the respective regions.

On the basis of the workshop, a set of guidelines were prepared on ways in which the management of production forests could be improved so as to favor the conservation of the biological diversity of these forests. These guidelines were subsequently sent for review to several international specialists in forest management and conservation. They constitute the main outcome of this project and are reproduced here in chapter 3.

The guidelines were submitted to the 10th session of the International Tropical Timber Council (ITTC) of the ITTO in Quito, Ecuador in May of 1991. The Council decided to constitute a working group to further develop the guidelines and to adapt them to the format of the existing ITTO "Guidelines for the Sustainable Management of Natural Tropical Forests" and "Guidelines for Establishment and Sustainable Management of Planted Tropical Forests". The process of drafting these definitive ITTO Guidelines is now well advanced and the Guidelines will be discussed by the ITTC at its 11th session in Yokohama in November and December.

Jeffrey A. Sayer
Gland, Switzerland
October, 1991
IUCN / ITTO

LIST OF CONTRIBUTORS

Jill Blockhus
Programme Assistant
IUCN – Forest Conservation Programme
Avenue du Mont Blanc
1196 Gland
SWITZERLAND

Charles Doumenge
Forest Conservation Officer
IUCN – Forest Conservation Programme
Avenue du Mont Blanc
1196 Gland
SWITZERLAND

Mark Dillenbeck
Forest Conservation Officer
IUCN – US
1400 16th Street, N.W.
Washington, DC 20036
USA

Jeffrey Sayer
Senior Advisor
IUCN – Forest Conservation Programme
Avenue du Mont Blanc
1196 Gland
SWITZERLAND

Ronald Kemp
UK Overseas Development Admin.
12 Westview Road
Warlingham, Surrey
CR6 JD9
UK

Per Wegge
Professor
Dept. of Biology and Nature Conservation
Agricultural University of Norway
P.O. Box 14
N-1432 AS-NLH, NORWAY

REGIONAL REVIEWERS:

Asia:

Sin Tuan Mok
7, Lorong Kemaris Dua
Bukit Bandar Raya
59 100 Kuala Lumpur
MALAYSIA

Latin America:

Alejandro Imbach
Senior Technical Advisor
IUCN – Central America
ORCA
Apartado 113
7170 CATIE
Turrialba
COSTA RICA

Gustavo Suarez de Freitas
Director Técnico
FPCN
Apartado 18-1393
Los Rosales #255
San Isadora
Lima 27
PERU

Africa:

Joseph Bawak Besong
Director Adjoint
Ministère de l'Agriculture
Département des Forêts
Yaoundé
CAMEROUN

François Wencelius
Forestry Specialist
Africa Technical Department
The World Bank
1818 H Street, N.W.
Washington, D.C. 20433, USA

COUNTRY STUDIES:

Asia:

Rabi B. Bista
Ministry of Forests and Soil Conservation
Babar Mahal
Kathmandu
NEPAL

Cesar Nuevo
Director, Institute of Forest Conservation
College of Forestry
University of the Philippines
Los Banos College (UPLB)
Laguna 3720
PHILIPPINES

Thang Kooi Chiew
Forestry Department
Jalan Mahameru
50660 Kuala Lumpur
MALAYSIA

Benni Sormin
School of Environmental Conservation Management
PO Box 5, Ciomas
Jalang Gunung Bata
Bogor 16001
INDONESIA

Latin America:

Clarence Bacchus
Managing Director
Trinidad&Tobago Forest Products
Ministry of Environment
TANTEAK, Whitehall
Port of Spain
TRINIDAD AND TOBAGO

Ivan Morales
CDC-Bolivia
C.P. 11250
La Paz
BOLIVIA

Julio Carrizosa Umana
Calle 10 No. 1-87
Apartado 60076
Bogotá D.F.
COLOMBIA

Eric Rodriguez
INRENARE
Apartado 2016
Paraíso
Ancón, PANAMA

José Flores Rodas
CATIE
Turrialba
COSTA RICA

Herbert Schubart
INPA
Alameda Cosme Ferreira
1756 CX P 478 - CEP 69083
Manaus, Amazonas
BRAZIL

Maria Marconi
CDC-Bolivia
Casilla Postal 11250
La Paz
BOLIVIA

Luis Suárez
Ecociencia
Avenida 12 de Octubre
PO Box 257
959 y Roca
Quito, ECUADOR

Edwin Mateo Molina
Director
CODEFOR-USAID Forestry Development Project
Honduran Corporation of Forestry Development
Apartado Postal no. 1378
Tegucigalpa, HONDURAS

Africa:

M. Mahamad Amine
Directeur Adjoint
Ministere du Tourisme
Yaoundé
CAMEROUN

Lumande Kasali
Département de l'Environnement et
Conservation de la Nature
14 Avenue des Cliniques
PB 12348 Kinshasa 1
ZAIRE

Jean Boniface Memvié
Directeur Général Adoint
 Eaux et Forêts
BP 2755
Libreville
GABON

Dominique N'Sosso
Conseiller au Ministre de l'Economie Forestière
BP 2153 Brazzaville
Républic Populaire du CONGO

Ounoh Nadjombe
Directeur de l'Environnment et du Tourism
BP 3114 Lome
TOGO

Kwabena Tufour
Director
Ghana Forestry Commission
PO Box M 434
Accra
GHANA

GLOSSARY OF ACRONYMS

CATIE	Center for Tropical Research and Investigation
CI	Conservation International
CGIAR	Consultative Group on International Agricultural Research
DP	"Domaine Protégé": State-owned forests which have no legal limitations on traditional rights and uses (Africa)
EIA	Environmental Impact Assessment
FAO	United Nations Food and Agriculture Organization
FINNIDA	Finnish International Development Agency
GF	Gazetted Forest
IBPGR	International Board of Plant Genetic Resources
IIED	International Institute for Environment and Development
IPCC	Intergovernmental Panel on Climate Change
ITTA	International Tropical Timber Agreement
ITTC	International Tropical Timber Council
ITTO	International Tropical Timber Organization
IUCN	International Union for Conservation of Nature and Natural Resources
NFAP	National Forest Action Plan
NGO	Non-governmental organization
PFE	Permanent Forest Estate: Includes both timber production forests and watershed protection forests (Malaysia and Indonesia)
PPF	Permanent Production Forest: Forests used for production of timber and/or non-timber products
TFAP	Tropical Forest Action Program (formerly Tropical Forestry Action Plan)
TPA	Totally Protected Area: encompasses IUCN categories I-V (see Asia overview chapter for definitions)
TMF	Tropical Moist Forest
UNCED	United Nations Conference on Environment and Development
UNEP	United Nations Environment Programme
Unesco	United Nations Educational, Scientific and Cultural Organization
WC	Working Circles: legal division of forest lands containing various specific management objectives
WCMC	World Conservation Monitoring Centre
WHO	World Health Organization
WRI	World Resources Institute
WR	Wildlife Reserve
WWF	World Wide Fund for Nature/ World Wildlife Fund

BIOLOGICAL CONSERVATION ISSUES IN FOREST MANAGEMENT

By
Jeffrey A. Sayer and Per Wegge

The loss and fragmentation of tropical moist forests is the single greatest threat to the world's biological diversity. The global significance of the phenomenon is recognised by the report of the World Commission on Environment and Development (the Brundtland report), which was adopted by the General Assembly of the United Nations. International action to conserve biodiversity is being harnessed under a "decade plan of action for conserving biological diversity," which is being prepared by the World Resources Institute (WRI), IUCN/The World Conservation Union, and United Nations Environment Programme (UNEP). An inter-governmental legal instrument for the conservation of biological diversity is being prepared by UNEP. Tropical Moist Forests are a major focus of all these plans, strategies and legal instruments.

We know that tropical forests are extremely rich in plant and animal species but the majority of these species have yet to be scientifically described and very little is known about their ecology. There is very little knowledge upon which to base conservation programmes but experience suggests that the retention of large areas representative of all forest types in a state of minimal disturbance must be the principal cornerstone of any conservation programme. National parks and equivalent protected areas are the main mechanism by which biological diversity conservation has been achieved. IUCN has prepared a number of studies of protected area priorities and action plans for their implementation. However, totally protected areas (TPAs) at present only cover about 5% of the tropical forest biome, and the rate of acquisition of new areas has declined markedly in recent years.

Pressures on land in the tropics are so great that options for establishing new totally protected areas are rapidly declining. It seems unlikely that it will be possible to allocate even as much as 10% of the moist forest biome to total protection. If natural or near-natural tropical moist forests were reduced to the 5%-10% of the biome that might be conserved in national parks and equivalent reserves, then a very large proportion of all tropical forest species will inevitably become extinct (Whitmore and Sayer 1991). There is, however, a broad consensus that it is possible to manage tropical forests in various ways for the production of timber and other products while still maintaining considerable biological diversity. Tropical forests are not static ecosystems maintaining a fixed climax species composition over a period of time. All forests have been subject to modification by climatic, geomorphological and human influences throughout their evolutionary history. Thus, whilst tropical forest lands may be extremely fragile when people attempt to convert them to other uses, the forests themselves are reasonably robust in their ability to recover from localised and periodic disturbance. Such disturbances caused by tree falls, storm damage and low intensity forms of shifting cultivation have contributed to the present pattern of diversity of the forests. Selective removal of a small volume of timber trees and the subsequent protection of the forest to allow regeneration of a further crop to be harvested after several decades constitutes a form of disturbance which might be expected to be compatible with the conservation of much of the biodiversity of the forests. It is known that many species, particularly larger mammals, are even favoured by the opening up of the canopy which occurs during forest exploitation. Several studies have also documented the rate at which animal and plant species recolonise forests which are regenerating after logging. These have in general shown that the speed with which original animal and plant communities are re-established is related to the degree to which the forest is disturbed and to the proximity of areas of undisturbed forests which can act as refuges for mobile species.

In many tropical countries, foresters have assumed that the maintenance of a permanent forest estate and the application of good forestry management practice would automatically lead to the retention of a forest environment which would support a majority of the original biodiversity of the area. It is only recently, with the acquisition of much greater knowledge of the true complexity of the forests, that it has been appreciated that even mild intervention may be inimicable to the persistence of some species. In particular, the importance of keystone species, those which play an important part in the life cycles of many other species, is now more fully appreciated. The selective removal of some of these species may cause a domino-like extinction of several other taxa. The precise nature of forest management systems applied to production forests will therefore have profound implications for the maintenance of some of these very complex communities of forest fauna and flora.

The objective of the present study is to evaluate the extent to which the future of the biological diversity of forests of ITTO member countries is secure. The national studies evaluate the extent and ecosystem coverage of protected areas and hence the degree to which these alone would ensure the maintenance of viable populations of all animal and plant species. But the main emphasis of the studies is to determine the extent to which forests managed for timber and other forest products can complement protected areas by favouring the maintenance of larger population sizes and more extensive ranges for forest species. Emphasis is therefore given to determining the legal and effective security of production forests, their spatial relationship to protected areas, the probable impacts of management systems on biological diversity and the overall extent to which production forests do satisfactorily enhance the biodiversity conservation functions of protected areas.

Results of the Study

The studies have shown that there are probably no countries where the coverage of protected areas is adequate to ensure the maintenance of all biological diversity. Usually the protected areas are not extensive and do not cover representative examples of all forest types. In other cases the areas of the protected units are too small or the management capacity too weak to ensure their integrity. Many protected areas are suffering encroachment from shifting cultivators or are subject to poaching for timber and wildlife products. There is therefore a general consensus that the objectives of biological diversity conservation can only be met if very significant additional areas of natural forests are retained under environmentally friendly systems of productive management.

Management for non-timber products is probably the use which is most compatible with biological diversity conservation, but forests managed for timber also have an important role to play. However, there is a consensus that the way in which production forests are managed at present is unsatisfactory and is inimical to the objectives of biological diversity conservation. The major issues are as follows:

1. There is great variation between countries in the extent of forests allocated for production forestry. Some countries, notably those in S.E. Asia, have allocated very significant areas as a legally gazetted permanent forest estate subject to management regimes which, if properly applied, would ensure the maintenance of a reasonable level of biological

diversity. In the case of Malaysia and Indonesia the permanent forest estate covers more than half of national territory. In Africa considerable areas of forest reserve have been established. Although, in general, management institutions are weak, the nature of the forest exploitation in Africa (highly selective) is such that the areas are not being significantly degraded and do maintain much of their biological diversity value. This is particularly true in the extensive forest areas of the Central African countries and less so in the largely deforested countries of West Africa. In South America, the concept of a permanent forest estate is less developed. Forest institutions are on the whole weaker and with the exception of Peru only rather small areas have any legal guarantees of security.

2. Few if any countries have specific measures in their forestry regulations to address the need to conserve the broad range of biological diversity in production forests. In general, management prescriptions aim to maintain timber yields, prevent soil erosion and protect hydrological functions. There are however some exceptions, a notable one being Malaysia where a system of Virgin Jungle Reserves exists within the permanent forest estate to conserve genetic resources and to provide sites for baseline scientific studies.

3. Virtually all the national studies indicate that the implementation of management prescriptions to ensure sustainability and integrity of the permanent forest estates is lacking. In the vast majority of cases, there are serious abuses of forestry laws, and in many countries, land which is designated for permanent forest cover is in reality being cleared for agriculture.

4. A very large proportion of the world's tropical timber production, and much of that which is traded internationally, comes from forests which are not legally allocated to the permanent forest estate. It is very difficult to get figures on the origins of timber but, at least in South America, most timber is obtained from areas which are being cleared for other uses.

5. In general, the condition of the permanent forest estates of countries is determined much more by the pressures to which they are subject than by any legal constraints upon their use which may exist. Thus, in remote areas with low population densities, forests usually remain in good condition even when they are subjected to logging. In countries where population pressures are higher, and particularly where demand for agricultural land is great, forests, even those within forest reserves, are often being seriously degraded.

6. In most countries forest products other than timber play an important role both in the local economies and for foreign export. Yet, prescriptions to integrate "other products" and multiple-use with timber production or biological diversity conservation in forest management schemes are virtually non-existent.

Conclusions

In spite of these serious reservations concerning the effectiveness of the existing forest management regimes, there is general agreement that sustainable natural forest management could have an important role to play in the conservation of biological diversity. Most countries are attempting to strengthen their ability to apply forest regulations and thereby ensure the sustainability of the management of significant areas of permanent forest estate. There is a general recognition of the need for international development assistance to be applied to help this process. The Tropical Forestry Action Plan is one mechanism to achieve this objective. There is also a consensus that the ITTO could also make valuable contributions both by helping countries to bring specific areas of forest under sustained yield management and by developing and disseminating technical information on management systems. Several country studies identify projects which ITTO could support to help achieve this objective.

ITTO has already taken several initiatives to promote sustainable forest management practices amongst its member countries. The "Guidelines for Sustainable Management of Natural Tropical Forests" adopted at the 8th session of the Council in Bali in 1990 were an important step in the quest for improved forest management. Also in Bali in 1990, the Council adopted the target that all internationally traded timber should come from sustainably managed forests by the year 2000. The general conclusion of the present study is that a significant proportion of this sustainably produced timber could be harvested from areas managed in a way which would favour biological diversity conservation. The prevalence of retail outlets for environmentally friendly products in the industrialised countries suggests that timber derived from such "conservation forests", might be particularly attractive to the consumer and thereby command a premium price.

This study recommends a set of guidelines for management of production forests which, if observed and monitored, would allow for recognition of a network of forest areas throughout the tropics managed in an exemplary fashion. These areas would constitute a tier of "partially protected areas" with a key role in supporting, extending and enhancing the biological diversity conservation role of national parks and other equivalent total reserves.

ITTO could promote the establishment of such a network of forests subject to exemplary management by taking the following initiatives:

1. Countries could be invited to nominate specific, legally defined, forest reserves for inclusion on a list of internationally recognised areas where forestry practice supports biological diversity conservation. ITTO could then conduct an independent evaluation of the integrity of the areas and their management regimes. If the results were acceptable, the area would then be included on the list. The process would be somewhat similar to that employed for the Ramsar Convention, which lists wetlands of international importance and the World Heritage Convention, which lists sites of global, natural or cultural significance.

2. Timber produced from listed sites would be labelled and might thereby be expected to have preferential access to markets in countries where consumers are concerned about the biological diversity of tropical forests. Consumers would be persuaded that by buying such labelled timber they would be contributing to good forest management and conservation whereas by buying unlabelled timber they might be contributing to forest destruction.

3. Sites on the list would be subject to periodic evaluation by ITTO and, in the event that management standards were found to have lapsed, they might be removed from the list.

4. ITTO might set itself a target of listing one hundred million hectares of such "approved" forests where sustained yield management practices are adequately applied, by the year 2000.

International recognition for such listings of sites could be given legal strength under the proposed global forest convention or charter. ITTO could then become the coordinating body for an important component of the convention or charter and a major force in the movement to conserve the world's tropical moist forests.

A NOTE ON THE DRAFT GUIDELINES FOR BIOLOGICAL DIVERSITY CONSERVATION IN FORESTS MANAGED FOR TIMBER

These draft guidelines were produced at a workshop at the IUCN General Assembly in Perth, Australia in December 1990. They were subsequently reviewed by several international specialists in forest management and nature conservation and the comments were incorporated into the draft. They were then submitted to all ITTO producer member countries for comments and these comments in turn have been incorporated into a final draft document.

This draft was presented at the ITTO council meeting in Quito, Ecuador in May 1991 where it was accepted as a product of the ITTO project entitled "Realistic Strategies for the Conservation of Biological Diversity in Tropical Forests". It was agreed that the draft would constitute a resource document for an ITTO working group which was established to prepare an "official" set of guidelines for approval by the ITTC at its meeting in Yokohama in November – December 1991.

The present version of these guidelines is substantively that which was presented in Quito and thus represents a consensus document based on contributions of the 200 people present at the Perth workshop, the ITTO member country specialists and a number of independent specialists. The changes that have been made since Quito constitute restructuring of the document for easier reference and facilitating the consideration of its contents in the process of elaborating the "official ITTO guidelines".

We would like to thank John Palmer, formerly of IUFRO in Vienna, the staff of the Forestry Department of FAO, Rome and Felix-Henri Maitre of the CTFT in Paris for their very significant contributions to the preparation of this document.

GUIDELINES FOR CONSERVING BIOLOGICAL DIVERSITY IN FORESTS MANAGED FOR TIMBER

BASIC PRINCIPLES:

1: Any disturbance of a forest, natural or man-induced, will alter it as a habitat for animal and plant species. Small-scale disturbances may enhance structural, floristic and faunistic diversity. Large-scale disturbances tend to simplify the ecosystem and result in loss of biological diversity.

2: The management of forests for timber production requires modification of the natural ecosystem to increase the yield of commercial species. Inevitably, some of the original forest species are then lost. Total aerial biodiversity may remain similar as other species colonise but because the colonisers are common and widespread and the displaced species are old-growth specialists, many with restricted ranges, the net result is a qualitative change in diversity favouring generalist species at the expense of old-growth specialists.

3: Detailed information on the ecology of all forest species and on their response to disturbance is not available. The safest strategy for conserving biological diversity is therefore to establish large undisturbed protected areas covering representative samples of all forest types.

4: Few countries are able to allocate sufficient areas to total protection to guarantee the preservation of all animal and plant species and their intraspecific genetic variation. In most countries, totally protected areas do not exceed 4-8% of national territory and individual areas are generally small. When species exist only in small isolated populations, they are susceptible to extinction caused by random environmental events and genetic deterioration. If natural forests are only retained in small isolated protected areas, then many species will inevitably be lost.

5: Forests managed for timber and/or non-timber products do provide habitats for many, in some cases the majority, of the plant and animal species found in pristine, unmanaged forests. The number of species persisting is dependent on a variety of factors, predominantly on the degree of intervention and modification of the original ecosystem. In some cases of low-intensity utilisation, forests may, at least in the short-term, have greater diversity than undisturbed ecosystems. Species composition will have changed, however, and some rare or specialised species may be lost.

6: Even though managed forests do not provide the requirements of all pristine forest species, they have a particularly important role in providing a buffering function around protected areas. This will especially benefit wide-ranging species, that are often able to persist in large production forests while they cannot persist solely in small isolated national parks and reserves. Because production forests are an extension of total forest area, although of a modified form, they will allow the existence of larger population sizes of such species than would be expected in protected areas alone. Production forests as buffer zones around natural forests will further act to minimise a severe problem in small protected areas.

RECOMMENDED ACTIONS:

1:   **Protected area systems should be established covering:**
- representative areas of all forest types
- examples of those forests having high species diversity or high levels of endemism
- forest habitats of rare and endangered species or species associations.

2:   **Protected areas should be as extensive as possible and should be linked by corridors of natural forest and surrounded by buffer zones of near-natural forest.**

PRINCIPLES RELATING TO THE RECONCILIATION OF CONSERVATION AND DEVELOPMENT:

7:     The macro-economic environment of many tropical countries is not conducive to sound resource management. In particular, debt repayment, terms of trade, communications and infrastructure policies all have important implications for forest conservation.

8:     Human population growth and increased demands on land, especially in the tropics, almost always mean that protected area systems can never be sufficiently extensive to adequately protect the full range of biological diversity of forests.

9:     The size and scale of human disturbance is increasing worldwide and there is a growing necessity for land to be commercially productive. Protected areas, while generating revenue from tourism and other low impact uses, will rarely generate as much revenue as timber production forests.

10:     The harvesting of non-wood products can often bring many benefits to forest-dwelling peoples and to national and regional economies and may be compatible with the retention of much of the biological diversity of the forest. Forests "reserved for extractive uses" may provide valuable complements to protected areas and should be part of any national forest land use allocation.

11:     A very large proportion of the world's forests are allocated for the production of timber and this situation is likely to persist. The future of much of the world's forest biological diversity depends upon the way in which these forests are managed.

12:     Theoretically, managed production forests often represent an attractive compromise between the need to use land to generate wealth and employment, thus helping to alleviate the poverty which drives people to clear forest for low-grade agriculture, and the desirability of conserving species. Production forests often represent a more tangible asset to societies in poor countries than do protected areas, and are more likely to be respected. In view of the worldwide decline in total forest area, even modified forests have an important part to play in preserving habitat, species and sub-specific diversity.

RECOMMENDED ACTIONS:

3: Decisions on macro-economic planning, export and trade policies, infrastructure and development assistance should take into account the impacts of these policies on forest biological diversity.

4: Forest management authorities should recognise that timber production forests have an important role to play in conserving biological diversity. Forest departments should have staff specialists in animal and plant ecology and this expertise should be applied to the preparation of forest management plans.

5: *Land tenure.* Long term security of the production forests should be guaranteed by legal gazettement, boundary marking and effective enforcement. Encroachment of logged-over forest by peasant farmers in search of agricultural land, conversion to non-forest use by speculators, and re-entry of loggers in response to changing demands for timber before the completion of adequate cycles of regeneration, must be prevented if the biodiversity of production forests is to be preserved.

6: *Forest-dwelling peoples.* Every effort should be made to involve local people in the management of the forests, and to ensure that they accrue benefits, which will motivate the people themselves to aid the maintenance of integrity of the forests. Local people's knowledge of forest resources and forest ecology can be valuable in developing sustainable forest management systems.

7: *Financial resource allocation.* Adequate financial resources should be available to forest managers for the effective management and protection of production forests. Timber royalty and stumpage charges must be adequate to cover forest management costs. These revenues must be applied to forest management activities, including those measures needed for biodiversity conservation.

PRINCIPLES RELATING TO MULTIPLE-USE, FOREST PRACTICE AND ENVIRONMENTAL SERVICES:

13: **Good forestry practice.** Strict application of the various measures which constitute "best forestry practice" (e.g., the ITTO Guidelines for the Sustainable Management of Natural Tropical Forests), especially minimising damage to residual trees, minimising soil compaction and disturbance from logging operations, yarding and road construction and maintenance, will in general favour biological diversity conservation.

14: **Protection of ecological functions.** Biological diversity conservation is favoured by measures to protect ecological functions. Hydrological functions are especially important and can be maintained by effective protection of all drainages and wetlands.

15: **Multiple-use and non-timber products.** In general, diversified production of a wide variety of products and services is more conducive to biodiversity conservation than managing the forest only for a limited number of timber species. Multiple-use management also generates direct benefits for local peoples. However, as with timber production, extraction levels must be regulated. Intensified management, which entails habitat modification, for any one product will reduce any gain derived from such diversification.

RECOMMENDED ACTIONS:

8:   **The ITTO Guidelines for the "Sustainable Management of Natural Tropical Forests" should be applied to all production forests.**

9:   **Undisturbed riparian forest strips should be established on each side of rivers and around water bodies. These should be a minimum of 20m wide along perennial streams of less than 20m width and 50m wide along larger streams, rivers and lakes.**

10:   **Significant areas of forest should be allocated for multiple-use management with biological diversity conservation as a major secondary management objective.**

FOREST MANAGEMENT PRINCIPLES:

16:   Certain forestry practices maintain the biodiversity conservation value of managed forests. These include the use of selective systems which take out small numbers of trees per unit area on a regular basis (such as those widely practiced in Africa and South-East Asia). These are often, though not always, more conducive to biological diversity conservation and to sustaining the functioning of the ecosystem than systems under which a larger proportion of trees are removed simultaneously during logging (such as those formerly used in the lowland Dipterocarp forests of Malaysia). Minimum intervention systems are more likely to sustain both timber yields and biodiversity values. Narrow strip cutting methods now being developed in Latin America (Peru) also seem to be compatible with biodiversity conservation.

17:   Animal and plant species will recolonise logged forest much more rapidly if small areas (approximately 100ha) of all forest types, covering in aggregate 10% of that forest, are protected as biological resource reserves or refugia (re: "virgin jungle reserves" in Malaysia). These should include riverine reserves and be in addition to forest areas which are inoperable for physical reasons, as these will usually represent different forest types.

18:   Many animal and plant species are adapted to old-growth forests. The survival of these species is favoured if some areas of forest are managed on very long logging-cycles. The value of areas of forest managed on long rotations will be greater if they are located so as to act as sources of colonists for intervening areas managed on shorter rotations.

19:   Animal and plant species cannot survive for long in isolated fragments of forest. Small isolated production forests and protected areas will have little value for biological diversity. The value for biological diversity of both categories of land will be greatly enhanced if they are located in large contiguous blocks. Internal zoning can be used to achieve the optimum balance between production and conservation objectives.

RECOMMENDED ACTIONS:

11: Biological diversity conservation needs should be taken into account in selecting silvicultural systems. Systems which minimise physical disturbance of the forest should be given preference.

12: Silvicultural systems which enhance natural regeneration of commercial species should be adopted. Hence, girth limits must be large enough to ensure adequate seeding and fruiting. Where needed, enrichment planting should use local wild seedlings, or seedlings raised from locally collected seed sources.

13: Refuge areas, with no extraction or other human disturbance, should be retained in logged forests. These should be as large as possible and should include all forest types in the locality. They should be retained through successive logging cycles.

14: Some forest areas should be managed under very long logging cycles so as to favour those species which are adapted to old-growth forest.

15: Forest areas allocated for logging and for conservation should be spatially distributed so as to avoid overall habitat fragmentation. Corridors of natural or near-natural forest should be maintained between forest blocks.

16: Silvicultural treatments should not include the suppression of species which are important in food chains or in providing ecological functions (keystone species). Some species may be important in the life cycle of commercial species as pollinators or dispersers and fruiting trees such as figs are important food sources for many forest animals. Use of pesticides, aboricides, herbicides and other chemicals should be kept to a minimum and manufacturers instructions for the use of each product should be strictly observed.

17: Trees with holes, standing dead trees (snags) and decomposing fallen trees are important in the ecology of many species and should not be removed from the forest.

18: The status of biodiversity conservation should be monitored at intervals by an appropriate census technique. Presence and relative abundance of indicator species should be recorded and if this suggests unacceptable reduction in biological diversity, then remedial action should be taken and future forestry operations modified.

PRINCIPLES RELATING TO SITES OF SPECIAL IMPORTANCE FOR BIOLOGICAL DIVERSITY:

20: Not all areas of production forest will have equal importance for biodiversity conservation. Sites of special importance for biodiversity will include:

- areas adjacent to protected areas
- areas with populations of rare or endangered species or with high concentrations of endemic species, or with exceptional species richness
- areas with unusual land-forms, geology, or other physical features not adequately represented in totally protected areas (TPAs)
- areas of forest type not represented in TPAs
- forest areas located so as to provide "corridors" between TPAs

21: In areas recognised as having special biological significance, a variety of supplementary management measures will be required. These may incur costs for timber operators, and this should be taken into consideration in the determination of concession fees and stumpage charges. Such losses of revenue can be justified on the grounds that they increase the biological diversity values of forests and help to maintain important ecological functions such as the preservation of water quality. Such special measures will be determined on a case-by-case basis in response to the specific conditions of the site. They may include:

- Lengthening of felling cycles, especially in areas adjacent to TPAs.
- Careful time-zoning of logging to ensure that areas at various stages of regeneration and mature stands are optimally distributed in relation to one another and to protected areas. In general, this means maintaining short distances between stands and forests of different ages, thereby preventing fragmentation of remaining old-growth forest blocks. It is important to avoid logging the whole of a particular forest type at one time.
- Strict regulation of log offtake per area. Lower offtakes will cause less canopy opening and less physical damage to the soil and vegetation.
- Generally greater vigilance in the application of all standard forestry practices. For instance, provisions should be made for independent, qualified personnel to supervise and monitor the field operations during pre-felling and harvesting operations.
- In particularly sensitive areas, it may be necessary to prohibit logging or to restrict offtake to a few high-value trees or species or to specified non-timber products whose extraction is consistent with the conservation objectives for the whole area.
- Special silvicultural measures may be used to enhance the value of forests for particular biological features. Timber operators may, for instance, be given incentives to selectively remove invasive exotic tree species (e.g. *Maesopsis eminii* which commonly invades logged-over forest in areas of East Africa outside of its natural geographical range).
- When adequate knowledge exists, it may be possible to identify special components of the habitat of rare or endangered or local species and protect them from disturbance (e.g. nesting sites, feeding concentration areas, lekking areas for birds of paradise etc).

RECOMMENDED ACTIONS:

19: When production forest areas are being allocated, and during preparation of forest management plans and concession agreements, areas of special significance for biodiversity conservation should be identified and delineated. This can best be accomplished in association with forest inventories.

20: The existence of areas of special significance for biological diversity conservation should be recorded and appropriate special conditions applied to logging, silvicultural treatments and post-logging protection. Additional uses of the forest area, such as harvesting of non-timber products and hunting, should also be regulated. An inter-disciplinary approach is required to determine the optimum management strategy whilst recognising biodiversity conservation as the primary objective.

21: Forest laws and management plans should allow for special restrictions and regulations to be applied to forestry operations in areas which are of special value for biological diversity conservation.

22: Special terms should be applied to forest exploitation agreements in those areas where biological diversity considerations require that forest operators incur higher costs than in equivalent forest areas in the same geographic area. These could take the form of reduced stumpage charges and similar taxes to offset greater exploitation costs, or opportunity costs incurred as a result of special restrictions on forestry operations.

ITTO AND THE CONSERVATION OF BIOLOGICAL DIVERSITY

by
Ronald H. Kemp

INTRODUCTION

The proposal to discuss the role of the ITTO in the conservation of biological diversity in the context of an IUCN workshop concerned with links between conservation and timber production arose from an initiative by the Government of Japan at the ITTO meeting in Bali, Indonesia, in May 1990. The proposal was approved by the member countries, two of which (the Netherlands and the United Kingdom) joined with Japan in providing financial support for the resulting ITTO initiative. The key contributions to the workshop discussions are those from the individual countries and regions from the participants in the meeting. However, in order to provide a common background of awareness of the nature of the ITTO interest and responsibilities, it was agreed to commission a résumé of past and current action, which forms the main part of this paper. Subsequently, the ITTO Secretariat has commissioned a consultancy aimed at the formulation of a 10-year action plan for the organisation in this field. The results of the IUCN workshop will therefore have the opportunity to contribute to the review of the issues and to the provision of information and advice towards the preparation of the draft action plan. While this paper should not attempt to anticipate the outcome of either the workshop or the consultancy study, it includes some suggestions on possible elements of future ITTO action, as an initial contribution to the workshop discussions.

ITTO's International Role

Origins and Organisation

The ITTO became effectively operational in 1987, at the time when international concern over deforestation, and particularly over the destruction of tropical rainforests, had begun rapidly to gather momentum. However the objectives, nature and structure of the organisation had an earlier and different origin, and some awareness of the historical setting is necessary to understand the present and possible future programme of ITTO in relation to the conservation of biological diversity.

The International Tropical Timber Agreement (ITTA, 1983) under which the ITTO operates is essentially a commodity agreement, arising from action taken by the United Nations Conference on Trade and Development in 1976, to initiate international negotiations on a number of individual products, one of which was tropical timber. Between May 1977 and June 1982, six preparatory meetings were held and two intergovernmental groups of experts were convened to examine research and development and market intelligence issues, which were seen as evident areas for possible ITTO action. Following a meeting on tropical timber in November 1982, the United Nations Conference on Tropical Timber was initiated in March 1983 and reconvened in November, when the text of the ITTA (1983) was finally established. Delays in signature and ratification by individual countries meant that the requirements for definitive entry into force were finally met on 31 March 1985, with 12 producing country members and 16 consuming countries, including the EEC and its member states.

The initial period of the agreement was set as five years, starting on 1 April 1985, and at its Sixth Session in May, 1989, the International Tropical Timber Council (ITTC) decided to extend the ITTA for a further two years, to 31 March 1992. This might be part of a four-year extension, to 31 March 1994, subject to further consideration at its Tenth Session, in 1991.

By the end of 1989, the ITTO had 46 members accounting for over 95 per cent of the world trade in tropical timber and nearly three quarters of the world's tropical forests. Since then, three more countries have joined. In common with other commodity agreements, the ITTA provides through the ITTO a forum for consultation and cooperation between its producing and its consuming country members, each group holding in total an equal number of votes. For producing members, each country's voting strength is related to the size of their tropical timber resources and timber exports. Consuming country votes are broadly proportional to their imports of tropical timber. The primary focus of interest of both groups is the expansion and diversification of trade and wood-based industries through sustainable utilisation of tropical forest resources, to achieve maximum economic benefit, and the equitable distribution of benefits between producers and consumers.

The highest authority in the ITTO is the council (ITTC), comprised of all the members, and served by three Permanent Committees:

(a)   Committee on Economic Information and Market Intelligence

(b)   Committee on Reforestation and Forest Management

(c)   Committee on Forest Industry

The Council and its Committees meet twice a year, alternately in its headquarters (Yokohama, Japan) and in a tropical member country (so far Brazil, Cote d'Ivoire, Indonesia and Ecuador).

Objectives and Operational Activities

The main objective is to provide an effective framework for cooperation and coordination in regard to all relevant aspects of the tropical timber economy, with a view to the expansion and diversification of international trade. The agreement includes specific reference to promotion and support for research and development, to improve forest management and wood utilisation; to the encouragement of reforestation and forest management; and to the development of national policies aimed at sustainable utilisation and conservation of tropical forests, of their genetic resources and of their contribution to maintaining ecological balance in the regions concerned.

The operational activities of the ITTO fall into two basic categories. One is comprised of project activity in the fields of research and development, market intelligence, further and increased processing, reforestation and forest management. The other, non-project, activities consist *inter alia*, of continuously monitoring trade and related activities, reviewing future needs of trade and the support and assistance being provided, identifying and considering problems and their possible solutions, conducting relevant studies and encouraging increased transfer of know-how and technical assistance. Of most direct interest in the conservation of biodiversity is the programme of the Committee on Reforestation and Forest Management. The functions of this Committee as set out in the agreement reflect the general objectives of the ITTO. However, it has been a high priority for the ITTO members to define more precisely the unique role of the organisation in this field of operation, in which other major international organisations and NGOs are already active. The results of this are summarised in later sections of this paper.

The ITTA makes specific reference to the need for coordination and harmonisation of the activities of the ITTO in the field of reforestation and forest management with those of other competent organisations, such as FAO, UNEP and the major development banks, in order to avoid duplication and to enhance complementarity and efficiency in the use of resources. The staff resources of the ITTO Secretariat are limited to nine professional staff, plus administrative assistance, whose responsibilities cover the full range of ITTA interests. The financial resources of the administrative account of the organisation, met by the annual contributions of member countries, are barely sufficient to cover the administrative budget (approximately US$2.8 million in 1990) at its current limited level of operation.

The ITTA established a Special Account to fund projects and pre-project activities. This is currently dependent mainly on voluntary contributions from member countries, most of which provide substantial and increasing support to international action for forest conservation and reforestation through other channels, bilateral or multilateral. In this connection, the definition of the precise role and comparative advantage of the ITTO in respect of action in regard to the conservation of biological diversity is critically important to the efficient use of international resources and to the achievement of international objectives in this field. IUCN, through its wide representation, has unique competence to assist in advising the ITTO on its future action in this regard.

The Conservation Challenge

The inclusion within the objectives of the ITTA of a conservation concern – to encourage the development of national policies aimed at sustainable utilisation and conservation of tropical forests and of their genetic resources – was a unique feature of this agreement as compared with other commodity agreements. It was therefore both far-sighted and highly significant. However, it remains only one among the eight objectives and was not foreseen as taking up a great deal of the time or financial resources of the ITTO, as compared with the "mainstream" activities concerned with aspects such as the promotion and diversification of international trade in tropical timber, improvements in timber marketing, distribution and market intelligence" the promotion of industrialisation, local processing, export earnings, wood utilisation, forest management, industrial tropical timber reforestation and so forth. For this reason, the size and nature of the ITTO Secretariat, and the composition of the organisation in terms of the representation of member countries in the Council and the Permanent Committees, were and remain illequipped to undertake the much greater and more active international role in the conservation of tropical forests now being thrust upon them.

This imbalance, between the expectation and demand for action from outside the organisation and capability within it, resulted from the rapid growth in public and media concern during the ten years or so between the original concept of the ITTA and its effective implementation. Latterly, the impetus to complete the drafting of the ITTA, and to secure its signature and ratification by the requisite minimum number of countries, was increasingly driven by the efforts of the international NGO community, notably the IIED, IUCN, WWF and Friends of the Earth, whose interests lay primarily, and in some cases exclusively, in the conservation objectives.

The complex issues involved were sharply revealed at the fourth meeting of the ITTC, held in Rio de Janeiro, in mid-1988. The venue for the meeting, in a country in which the subject of deforestation is linked in the minds of both the public and the media not only with major environmental issues, but with concern over the impact on forest-dwelling peoples, raised expectations for ITTO action some of which were clearly beyond the declared objectives of the organisation while others implicitly involved a conflict between its conservation objective and those more directly concerned with the international trade in tropical timber.

One of the earliest actions of ITTO in relation to its conservation objectives was to commission a study, carried out by the IIED, into the sustainability of current practices in natural forest management for timber production in its member countries. This study, which was completed in September, 1988, and presented to the Third Session of the Permanent Committee in November, 1988, was subsequently summarised in the book "No timber without trees" (Poore, *et al.*, 1989). The results of the survey undertaken in Africa, Asia, South America and the Caribbean (some 20 countries) revealed that only or very small fraction of the tropical moist forest was demonstrably under sustainable management in the fullest sense. At the same time, the report was careful to point out that in many areas some elements of sustainable management were being practised and that overall the inadequacies were the result of failure to apply the available techniques of regeneration and management with sufficient rigour and consistency, rather than the lack of understanding of what should and could be done.

Nevertheless, the conclusion was clear that current tropical timber production from natural forests was contributing to depletion of the forests, and that unless and until the necessary improvements in management practices were introduced, there must be a danger that ITTO action to promote the expansion of international trade in tropical timber would conflict with its objectives for the conservation of tropical forests and their genetic resources. For those whose primary concern was the conservation of the rapidly diminishing tropical rainforests, therefore, the implications for ITTO action were to reduce or halt the logging pending the introduction of effective sustainable management, for example through the imposition of restrictions on the importation of tropical hardwood into the consuming countries. Such action would clearly have an adverse impact on national economies, timber exporting enterprises and individuals in the producing countries, and therefore run contrary to these areas of ITTO interest.

Producer Interests in ITTO

Producer country delegations to the ITTO have been generally supportive of the strong emphasis given so far to conservation issues particularly by environmental interests among the NGO community and the media, and to related reforestation and forest management activities. This attitude is due at least in part to the relatively strong representation of the "growing side" of the forestry profession in producer country delegations, as compared to consumer country delegations, in which trade and industrial expertise and interests have been more strongly represented. Almost without exception historically, the forestry sector in tropical countries has been starved of the resources needed for proper management and regeneration of the forests. ITTO was seen by foresters in the producing countries as a new vehicle through which to secure urgently needed financial resources, particularly by transfer from the industrialised consumer countries, to support sustainable management and conservation of tropical forest resources. In their view, lack of adequately trained staff and other investment in the sector was the principal cause of inadequate management. Sustainable utilisation and conservation of forests and of their genetic resources are inevitably more costly than short-sighted and wasteful logging practices. They are only achievable given adequate investment in staff and infrastructure, and in the husbandry of the forest capital rather than its rapid exploitation to produce more immediate material and financial benefits. Given the other, and in many respects much greater, potentially destructive pressures on tropical forests and forest lands, from agricultural expansion and other demands of expanding populations, any restriction on the export of timber and timber products and therefore on the evident value of the forests might lead to their more rapid destruction. Conversely, there is increasing recognition internationally that deforestation will be stopped only when the natural forest is seen to be economically more valuable than alternative uses for the same land. Such economic judgements must attempt to take account of the long-term benefits of conservation, but cannot ignore the short-term financial costs to tropical countries already under severe economic constraints. (Kemp 1990)

Consumer Interests in ITTO

The industrialised consuming countries in ITTO, during the long gestation period of the ITTA had seen the organisation primarily as a forum for consultation and cooperation through which to promote the expansion and diversification of international trade and market transparency, and to support associated research and development activities, rather than a major new executing or funding mechanism. The agreement also foresaw the use of the ITTO as a forum in which to encourage members to support and develop industrial tropical timber reforestation and forest management activities, as well as national policies aimed at sustainable utilisation and conservation objectives. However, the ITTO was not designed to administer the transfer of substantial financial resources for large scale reforestation and forest management activities, more appropriate for international funding through the major development banks or other multilateral and bilateral aid organisations. To a large extent, therefore, the resolution of the possible conflict of interest between producer countries on the one hand, concerned to maintain and increase their revenue from forest exploitation, and conservation interests on the other, in both producer and consumer countries, must be dependent on international action outside as well as within the organisation, to provide the necessary means and financial resources to achieve sustainable management. Nevertheless, it is also clearly central to ITTO's objectives and mandate to promote such action on the part of member countries and other international organisations through its role as an international forum.

The great majority of the public and the media in the consumer countries who are aware of the existence of the ITTO are concerned above all with its possible role in the conservation of tropical rainforests, and only secondarily, if at all, in its role in promoting the sustainability of trade in tropical timbers, which many of them may see as being contrary to conservation objectives. It is arguable that, despite the commodity focus of the ITTO, the interests of the great majority of the people in the industrialised timber consuming countries and their national economic interests, lie primarily in the organisations role in the conservation of the forests and their included biological diversity, and only secondarily in their role as a source of timber. Although there would be considerable short-term disruption of trade and industrial practices as a result of any substantial reduction in tropical timber supplies, alternative sources or substitute materials could be obtained. In the last resort, the cessation of international trade in tropical timber would have far less impact on the national economies of the importing countries than on those of the countries of origin. (Kemp 1990)

The Global Interest

The increased international interest in the fate of tropical forests is primarily related to their environmental values, particularly through the supposed link between deforestation (or conversely forest conservation and reforestation) and the impact of the "greenhouse effect" on global and regional climate. The issues involved as well as the considerable degree of uncertainty that exists over the rate, extent, likely effects and even the very fact of global climate change, have been so widely discussed as to need no repetition here, (IPCC 1990). Whatever the uncertainties, the probable cost of failing to respond adequately and in time to the possibility of significant climate change gives special importance to response strategies which will provide substantial benefits in their own right, independently of their beneficial influence on global climatic stability. International concern over the adverse environmental and economic impacts of deforestation at local, national and regional levels in tropical countries, and over the resulting poverty, unemployment, migration and social instability, is already securing some increased international aid to the sector. In addition, the importance of the tropical moist forests as major reservoirs of biological diversity is gaining increased attention. (McNeely *et al.*, 1990).

Whatever the uncertainties of the economic value of tropical forests in terms of their included biodiversity, the certainty of irreversible loss of a significant quantity of the existing genetic resources through current rates and patterns of deforestation is beyond question. In this respect, the global significance of the tropical forests in terms of their biodiversity is even greater than their role in global climatic stability. Given the political will, other options to reduce the "greenhouse effect" directly, for example through reduction in industrial gas emissions, are open to the industrialised countries, just as alternatives could be found for tropical hardwoods in both industrial and domestic applications in the importing countries. However, there is no realistic alternative means to the conservation of the main body of the biological diversity of the tropical forests apart from the conservation of the forest ecosystems.

To the extent, therefore, that conservation of tropical moist forests may be linked to their sustainable management for the production of timber and other products, the ITTO should have a significant role in the conservation of biological diversity, in concert with other appropriate international organisations.

The Role of Other International Organisations

Whereas the ITTO has special responsibilities in regard to tropical timber supplies and timber-producing forests, with associated interest in the conservation of forest genetic resources, the leading role in international action for the conservation of biological diversity lies with other organisations either within the United Nations group of agencies (e.g. FAO, UNEP, UNESCO, WHO etc) or outside (e.g. IUCN, WWF, IBPGR etc). FAO, UNESCO, UNEP and IUCN coordinate their work in the environmental field through the Ecosystem Conservation Group, under whose aegis a working Group on *in situ* conservation of plant genetic resources was established in 1984. The role of international organisations in the conservation of plant genetic resources was reviewed in a paper to the Symposium on the Conservation of Genetic Diversity held in Davis, California, USA in July 1988 (Palmberg and Esquinas-Alcazar, 1988) and the activities of the principal agencies concerned in this field are well known to IUCN members through recent meetings and current literature (e.g. McNeely *et al.*, 1990). There is no need in this meeting to restate IUCN's international role in this field. However, it may be helpful to consider how activities of other major international organisations relate to ITTO responsibilities and action programmes bearing on biodiversity conservation.

FAO first directed attention to the role of plant genetic resources in regard to food and agricultural development over forty years ago and its Panel of Experts on Plant Exploration and Introduction was established in 1962. The FAO Panel of Experts on Forest Gene Resources was established in 1968 and had its Seventh Session in December 1989. The Forestry Department of FAO provides the secretariat to this panel and in conjunction with it helps to coordinate international action in the exploration, collection, evaluation, conservation and utilisation of forest genetic resources at the global level. Initially, the focus of attention of the panel was towards plantation forestry, including both *in situ* and *ex situ* conservation, and on technical issues related to the evaluation and improvement of genetic materials, particularly for reforestation programmes, which are still central to its concerns. However, the report of the Seventh Session calls for special attention to the development of methodologies and pilot activities in *in situ* conservation, including its incorporation as an integral part of the management of forests for other purposes. The panel requested FAO to continue to collaborate in national and international efforts to develop practical methodologies to define, assess, evaluate and manage biodiversity, and called for appropriate allocation of resources to meet the needs of conservation in the context of sustainable development.

The Forestry Resources Development Branch of the FAO Forestry Department, which is responsible for these issues related to the conservation of plant genetic resources, is also the focal point within the organisation for the assessment, monitoring and management of forests and woodlands. These are fields of action in which FAO has had exceptional experience and their links to the conservation of biodiversity are of evident importance both to the FAO and ITTO.

UNESCO has played a leading role in the creation of essential protected areas through its Man and the Biosphere programme, and through the network of Biosphere Reserves. There is now increasing recognition within UNESCO and among other organisations which have led the international action towards the creation of protected areas dedicated to the conservation of biological diversity that reliance on such areas alone cannot ensure adequate protection of genetic resources. Both UNESCO and IUCN are encouraging international consideration of the approaches to extended action to conserve biodiversity through the incorporation of this objective in areas managed for simultaneous production of other goods and benefits, including timber. The scientific studies needed to understand the dynamics of tropical forest ecosystems can be complex but the objective of such studies is clearly closely allied to ITTO interests in reconciling the needs of conservation with its responsibilities for the promotion of sustainable timber production.

UNESCO also plays a key role in the examination of the adequacy of worldwide coverage of representative samples of ecosystems in Biosphere Reserves. The activities of the IUCN/WWF/UNEP-sponsored World Conservation Monitoring Centre (WCMC), particularly in conjunction with the FAO assessment of global forest resources, are of key importance in assessing the extent to which the existing system of reserves is likely to meet the need for conservation, both in terms of their location and of their effective management and control.

Biodiversity conservation is among the key environmental issues under the international mandate of UNEP, which has provided significant support to a large number of projects and activities, in conjunction with FAO and other appropriate UN agencies, including the development of models for field application, for example through the FAO/UNEP project on *in situ* conservation of forest genetic resources, involving the establishment of pilot areas in three continents. UNEP also collaborates closely with IUCN and other non-UN agencies in these and related issues.

WWF-International and its principal national subsidiaries (e.g. WWF-UK and WWF-USA etc) support a very wide spectrum of activities related to the conservation of biodiversity in many countries. These include substantial projects in association with national governments and official aid organisations focussed on the sustainable management of areas of natural forest and surrounding lands for the simultaneous conservation and production objectives referred to earlier as of key interest in the context of the ITTA. Although such approaches are mostly at an early stage of exploration and development, it is already clear that social considerations and expertise, related to the needs and activities of local people in and around the forests, as well as the biological and physical sciences associated with land use and natural resource management, are of key importance to their success.

Of fundamental importance in all conservation action are the interests and priorities of the national governments as well as the local communities concerned. In addition to the international organisations referred to above many national organisations, both official and non-governmental, often assisted by multilateral or bilateral aid organisations and development banks, are now involved in the attempt to conserve specific areas or elements of biological diversity. Given the complexity of the ecosystems and the lack of adequately trained taxonomists, ecologists and other scientists in the countries concerned, the provision of technical cooperation as well as financial support is critically important. The Tropical Forestry Action Programme (TFAP) is still in principle a valuable mechanism for coordinating international action in this field in accordance with national objectives and as an essential part of sustainable development programmes.

Given the wide spectrum of scientific information that must be urgently gathered to guide the location and management of conservation programmes, international research organisations such a IUFRO also have an important role, in concert with national research centres. The urgent need to strengthen international research in forestry generally, with due attention to genetic resources, is under current examination by the CGIAR, drawing on the guidance provided by an international task force which reviewed the needs for intensified research in forestry and agroforestry. It seems likely that the IBPGR and possibly IFAR will play an increased role in the exploration and evaluation of forest genetic resources in future, particularly through the provision of information needed for action to conserve selected populations and genotypes *ex situ*.

Many of these on-going or projected international actions relate closely to the interests of ITTO and to its conservation-related objectives. It is essential for the organisation to keep itself fully informed of relevant activities in order to avoid possibly wasteful duplication of action and to take full advantage of the available information and other action relating to the link between tropical forest management and biodiversity conservation.

ITTO's Emergent Action Programmes

Consideration of the responsibilities and competence of the ITTO relative to those of other international organisations has been an important factor in the development of appropriate action programmes for the organisation and its three permanent Committees. In view of the very broad and complex objectives set in the ITTA, the council, at its Third Session in November 1987, requested a document to outline priority areas and criteria for programme development and project work. The resulting paper, prepared by the ITTO Secretariat, took account not only of the relevant articles of the ITTA and other related documents, but also of views and ideas expressed by outside and non-governmental organisations, notably WWF, and of proposals for collaboration with UN agencies and with the work of the TFAP. Two key statements in the Secretariat paper were firstly that an analysis of the objectives of the ITTO could be reduced to the single issue of how to conserve tropical forests by using them for timber exports, and secondly, the related thesis that the only realistic approach is to make forest lands more valuable under sustainably managed forests than the alternative uses to which they are being diverted. While such general principles were accepted, it was decided that more detailed work was needed to provide clear action plans and work programmes and, given the very limited capability of the Secretariat, this could best be achieved through expert panels and working groups drawn from the member countries and organisations. This procedure was initiated in the Permanent Committee on Reforestation and Forest Management, with the nomination of an expert panel, including representatives of both producer and consumer member countries, FAO, environmental NGOs and the ITTO Secretariat. The panel started work in December 1988 and submitted its report to the Sixth Session of the Council, in May 1989.

Subsequently, similar action has been taken in respect of the action plans and work programmes of the other two Permanent Committees, and it is significant that both these relevant Committee reports, while primarily concerned with the "mainstream" issues related to forest industry and to economic information and market intelligence respectively, emphasise the need to situate the trade in tropical timber in its wider development context of environmental considerations and sustainable utilisation, including the implications for the conservation of the forests and their genetic resources.

A most significant and far-reaching strategic conclusion of the ITTO to emerge from these considerations at the Eighth Session of the Council, in May 1990, was the adoption of the objective that all exports of tropical timber should come from sustainably managed forests by the end of the present century, in the year 2000. Insofar as the concept of sustainability must include the conservation of genetic resources this target demands urgent consideration of ITTO's possible role in relation to biological diversity.

The Eighth Session of the Council also requested the integration of the action plans of the three permanent committees into an overall and comprehensive ITTO Action Plan, to be considered at its Ninth Session in November, 1990. Given the content of the respective Committee reports, it can safely be assumed that the overall Action Plan will give appropriate emphasis to conservation issues, including those relating to ITTO's role in the conservation of biological diversity. While elements of this role (e.g. in regard to the use of lesser-known species, and the implications for forest management plans of any major changes in the range of species harvested) are reflected in the action programmes of all three Committees, it is primarily the programme of the Reforestation and Forest Management Committee which will determine the ITTO's role in biodiversity conservation.

The current action programme of this committee contains proposals for action in eight major areas. These are listed below, with a brief review of the possible relationship of each to the conservation of biological diversity.

(i)   **Guidelines for "best practice" and sustainability in the sustainable management of tropical forests.**

The concept of sustainable management is extremely wide, embracing not only sustained yield of timber, with due attention to the quality, variety and economic value as well as volume, but also the range of other goods and benefits from the forest, including its environmental and ecological influences. The primary and subsidiary objectives of management, as well as the local conditions, both ecological and socio-economic, which govern the options open to the manager, vary from one forest area to another. Nevertheless, to achieve the objectives set, requires precise guidance, rules, codes of practice and manuals of procedure at all operational levels. Moreover, all guidelines are valueless if not effectively applied in practice. It follows from all of this that any guidelines must be produced within, and under the direct guidance, of each producing country and tailored to specific objectives and conditions, while at the same time incorporating the essential principles of sustainable management.

Under its 1990 Programme of Work, the Council appointed a small working group representative of both producer and consumer countries, and of international and nongovernmental organisations, and the timber trade, to develop the necessary principles and recommendations for action by ITTO to institute internationally-accepted guidelines. The report of the Working Group to the Eighth Session of the Council in May 1990, resulted in the adoption of an agreed set of principles, and associated actions, formally endorsed by all members as an international reference standard for the development of more specific guidelines at the national (and subsidiary) levels in all timber producing countries. The principles agreed include specific references to the setting aside of land for native conservation, to the incorporation of objectives such as the conservation of species and ecosystems in schemes of forest management primarily for timber production, and to the avoidance of habitat disruption and reduction of biological diversity in harvesting operations.

The adoption of this set of principles by the ITTO is given additional significance in the context of the declared target date for all tropical timber exports to be derived from forests under sustainable management by the year 2000. However, effective compliance still depends on the adoption and implementation of national guidelines in each producer country. Moreover, unless such national action is taken, the ITTO guidelines will lose international credibility.

(ii) **Develop the economic case for natural forest management.**

This activity is central to ITTO's interest in reconciling the conservation of tropical forest resources with their utilisation for timber production. It embraces the full evaluation of all benefits from the forest as well as the costs of management to achieve them sustainably. The subject is under current examination and debate in other international organisations, including the major development banks and UN agencies, as well as nongovernmental organisations, many of which have greater internal resources to devote to such questions. Two ITTO-sponsored studies on "Multiple Use Management of Tropical Forests" and "Natural Forest management for Sustainable Timber Production", contained in ITTO documents PPR 7/88 (F) and PPR 11/88 (F) respectively, drew attention to the very limited data available and the needs for further research to be incorporated into appropriate ITTO projects. The need for close cooperation with FAO and with national research institutes and technical assistance agencies was emphasised by the Permanent Committee. As a specific ITTO contribution to the collection and collation of the necessary basic data, the organisation decided to support action to gather such information initially in respect of one region (Asia Pacific) to be coordinated by the Forest Research Institute of Malaysia. Subject to progress and results in this study similar action may be initiated in the other major tropical regions.

(iii) **Strengthen policy initiatives embracing the forestry sector.**

This is seen as a central and continuing activity of ITTO in its role as a forum both in the Committees and in the Council, which will also be assisted through specific actions such as the adoption of the international principles for sustainable management and their incorporation in national practices, including the linking of forest policies to broader government policies involving other sectors. These must include the preparation and implementation of national conservation strategies and attention to related land tenure and land-use issues.

(iv) **Increase awareness and mobilise support to ensure the sustainable management and conservation of tropical forests.**

International awareness of the need for conservation is now strong and ITTO attaches interest to establishing a better understanding of the links between conservation and sustainable use of the forests. This is dependent on other action to gather relevant economic data and to develop demonstration models.

(v) **Develop demonstration models of management for the sustainable production of timber and non-timber products and conservation.**

The objective is to establish an international network of sites for both demonstration and training purposes, drawing both an existing example of successful forest management and on others to be established through ITTO-sponsored projects designed for that purpose. Examples of the latter are being established in Brazil, Bolivia, Malaysia and elsewhere. Of particular interest in this field of action will be models to incorporate the conservation of genetic resources and biological diversity generally within management systems for the production of timber and other products. This implies the need for coordination of action with other international agencies active in this field, notably FAO, UNESCO and IUCN.

(vi) Strengthen research on responses to silvicultural treatment.

Several ITTO-funded projects have been designed to study aspects of the responses of forests to silvicultural treatment and to develop research capabilities in this field. The influence of stand treatment on the recruitment and growth of the next timber crop, and on related forest dynamics, is of course a long-established field of research among both national and major international agencies. ITTO has a particular interest in encouraging the efficient coordination of international action in this field. Insofar as sustainable management is seen to include conservation of biological diversity there is also an ITTO interest in examining the effects of silvicultural treatments in this respect, for example in the demonstration areas referred to above.

(vii) Develop human resources in tropical forest management.

The lack of sufficient and adequately trained staff is a major constraint on sustainable management and ITTO is contributing to the total international effort, strongly developed in all major multilateral and bilateral aid programmes, to increase the capability in the tropical countries through human resource development. However, ITTO has no specific expertise in regard to training for the conservation of biological diversity.

(viii) Consider incentives to encourage sustainable management.

This was the subject of an ITTO seminar on the occasion of the Eighth Council Session in Bali, Indonesia, in May 1990, and is being studied under an ITTO preproject whose final report is due to be presented in 1991. There are no specific implications for ITTO's role in the conservation of biological diversity, other than through the establishment of sustainable forest management.

Some elements of future ITTO action

The action needed for conservation of the biological diversity in tropical forests is on the one hand a major intensification of scientific research and technical cooperation programmes, to collect the essential information, through surveys, inventories, ecological studies etc., and on the other hand, the allocation of financial resources both to support this scientific action and to meet the operational costs of effective conservation. The latter will be substantial, both in terms of direct costs of management and the indirect costs of opportunities foregone in the setting aside of substantial areas of land and forest, which might otherwise have realised immediate short-term revenue, food production, energy supplies etc. through conversion of the land to other uses. Although ITTO is neither in itself a major source of relevant scientific expertise, nor of financial support, it could nevertheless play a significant and perhaps even decisive role in international action, based on its position as an international forum concerned with the links between the productive use and conservation of the forest resources. This is dependent on obtaining widespread recognition of the possibility, both technically and in financial terms, to combine management for local and national socio-economic benefit with international conservation interests, and widespread acceptance of the changes needed to realise this possibility. These must include increased investment in the sector and secure financial provision to meet the costs of management, preferably based on an appropriate level of revenue through the market. This implies substantial changes in current patterns of international trade and industrial investment to achieve the combined national and international objectives.

Project Related Activities

Within the current ITTO action programme, several activities listed above (para 4.5), related for example to (i) the development of national guidelines for sustainable forest management (ii) development of the economic case for natural forest management (v) the establishment of demonstration models for the sustainable production of timber and non-timber products and conservation (vi) research on responses to silvicultural treatment; and (viii) the exploration of incentives to encourage sustainable management, could be pursued through ITTO-sponsored projects. Activities of other ITTO Committees related for example to the use of lesser-known species, may also have conservation implications and be furthered through projects aimed at deriving the necessary information to assist productive sustainable use of the forests.

To secure the necessary scientific expertise in the most efficient manner, close cooperation will be essential with both national and other international organisations also active in this field. The ITTO-sponsored model demonstration areas (para 4.5 (v) above) should provide excellent opportunities for such coordinated and cooperative research into the incorporation of biodiversity conservation within management systems for the production of timber and other forest products. These demonstration areas would be monitored through interventions at various stages from initial inventory through harvesting regimes and silvicultural treatments. Such models may also involve the use of artificial regeneration, either for enrichment or to establish blocks of plantation in large deforested gaps or as "buffer" strips to protect the natural forest from further encroachment, fire etc. Although such activities have only indirect influence on the conservation of biological diversity, they may be linked to action to conserve the genetic resources of selected timber species. This is a field of action already strongly developed in other international organisations, such as FAO, and supported by bilateral aid programmes and by the private sector. It is also a candidate for increased support through the CGIAR and IBPGR, and therefore less urgently in need of ITTO support.

The fundamental areas of research needed to secure the information for biodiversity conservation are especially in taxonomy, autecology and the dynamics of the ecosystems in relation to the pressures and changes being imposed. The resources needed to undertake thorough and systematic exploration and documentation of the tropical forest are clearly impossible to achieve or to justify in economic terms for their academic and scientific interest alone. However, much more effective use could be made of the opportunities offered by management-oriented studies (e.g. forest inventories primarily concerned with the population and population-dynamics of timber trees) to gather information on other aspects of biological diversity in the forests. For example, forest inventory and survey teams in a given area can be joined at little additional cost by taxonomists or ecologists from local scientific institutes, universities or international organisations, to gather data on other flora and fauna. What is required to achieve this is deliberate planning in advance and good communications between the organisations concerned. Above all, it requires government authorities, timber concessionaires, trade interests and conservation groups to work together in close collaboration. Such arrangements could be achieved in the context of national guidelines for sustainable management, related to the ITTO standards (para 4.5 (i) above) and action to ensure their effective implementation could be written into the management plans at appropriate regional and local levels. This should take account of available information on the conservation status of tropical timber species in international trade, which is the subject of a current ITTO-sponsored study at the WCMC.

Ultimately the effective application of conservation objectives and programmes is dependent on the interest and involvement of the local human populations in and around the forest. Unless the right incentives exist to secure their positive involvement, no amount of legislation or scientific argument can be effective. To a considerable extent, this aspect can be examined under project activities related to the economic case for natural forest management (para 4.5 (ii) above), to increasing awareness and mobilising support (para 4.5 (iv) above) or in connection with the demonstration models referred to earlier. In addition, some related project activity may emerge from the on going pre-project study of incentives to encourage sustainable management (para 4.5 (viii) above). Appropriate involvement of local communities should be incorporated in the national guidelines for sustainable management.

Non-Project Action

Scientific and technical cooperation programmes to develop the necessary information, methodologies, demonstration and training activities to assist conservation of biodiversity will help establish the economic case for the required investment to cover the costs of protection and sustainable management of the forests. Already there is considerable international acceptance of the need for the industrialised countries to contribute substantially more to the costs of conservation of tropical forests, taking account of their global environmental values in terms of climatic stability and genetic resources. Among the ideas put forward are various aspects of financial payments or other resource transfers from rich to poorer countries to be made either as a single down-payment or at regular intervals for an indefinite number of years, as both incentive and assistance for the conservation of the forests. These include "debt-for-nature swaps" as well as the mobilisation of funds dedicated to forest conservation and reforestation, possibly related to a levy on international trade in tropical timber. The proposed Global Environmental Facility being prepared by the World Bank and the UNDP might be a possible source of funds. Any arrangement must seek to secure not only the protection of a given selected area in perpetuity but also appropriate management to conserve the genetic resources within it. This means that it must survive through successive government changes and political/economic crises, and in the face of increasing population pressures and demands for land, which could lead to slow attrition or sudden reversal of the agreed conservation objectives, for short-term local benefit. The capability of the forest to produce tangible and sustainable benefits for the national economy and local populations simultaneously with serving its conservation objectives is the best guarantee of long-term security.

The ITTO action in agreeing both the international standards for sustainable management and a target date by which to achieve this in respect of international trade in tropical timber, as well as the examination of incentives to sustainable management, provides a basis for further action to secure the financial resources needed in both the short and the long-term. The immediate need is to establish all the conditions for sustainable management in the countries concerned, namely investment in infrastructure and skilled manpower for efficient management, harvesting, processing and marketing of the forest resources. This will require very substantial initial investment, for example through official aid programmes, both bilateral and multilateral, and through the private sector, including the possible use of "debt-for-equity swaps" to assist the establishment of appropriate local processing. The possibility of a levy on timber in international trade to assist in meeting the costs of sustainable management and reforestation is among the ideas which have been proposed but have so far had insufficient attention either within ITTO or elsewhere. However, the essential aspect for ITTO is the need to review and revise the existing patterns of international trade and investment, to ensure the appropriate levels and distribution of financial benefits in the producing countries to secure the positive link between international trade and sustainable management, including conservation of objectives.

The United Nations Conference on Environment and Development (UNCED), in 1992, will provide an unprecedented and final opportunity for international action at the level and scale needed to achieve the major reformation of existing practices that will bring the global interest in tropical forest conservation, and the national interests of tropical countries in the productive values of their forests, to common focus in sustainable forest management. This will require radical and far-reaching reform of international agreements regarding debt management, trade regulation and the provision of development assistance, as indicated earlier, to link enhanced and sustainable revenue from international trade to the conservation of the forest resources and their biological diversity. The final months leading to the UNCED meeting are critically important. They should be a period of intense expert examination of the complex issues, leading to the provision of coherent, clear advice at a high political level in all participating countries, and in relevant international organisations, that will ensure a basis for decisive action during the conference itself.

The Preparatory Committee for UNCED, at its first session in August 1990, took note of the interests of timber-exporting countries in their forests as a source of timber, as well as in terms of their multiple uses and as habitats of biological diversity. It also noted the ITTO's interest and competence in regard to some of these issues, and encouraged increased ITTO involvement in the discussions related to the preparations for UNCED, including the consideration of possible international conventions or other legal instruments concerned with biological diversity, global climate and the forests themselves.

The role of ITTO as an international forum, and the source of information and advice in these critically important preparatory discussions, must have the highest priority. This begs the question of the extent to which the resources of the ITTO Secretariat are adequate to allow its appropriate contribution to the ongoing international action in these conservation-related fields whose significance, both to ITTO itself and internationally, were not foreseen when the organisation was designed. However, a significant feature of ITTO which was apparent both in the action which led to the establishment of the organisation, and in its subsequent development, has been the involvement of the environmental NGOs both in their respective countries and within ITTO itself. The influence of the NGOs in the preparations for UNCED may well be decisive in determining whether the conference leads to correct and decisive action, or to further procrastination.

Despite the real complexities involved in the attempt to achieve the necessary positive and enduring links between the continued and enhanced productive use of tropical forests in international trade, and the conservation of their biological and genetic resources, there are three simple underlying and interrelated requirements. These are the following:

(i)   The transfer of resources from the rich, consuming countries to the tropical producing countries sufficient to meet the additional costs of sustainable forest management, including provision for the conservation of biological diversity.

(ii)   The adoption and strict application in each tropical producing country of national guidelines for sustainable forest management which ensure that the resources transferred through the international market, or by other means, are properly and equitably distributed among the agencies and communities whose cooperative actions are needed to achieve the objectives of sustainable management.

(iii)   The recognition by all influential participants in the UNCED meeting, and the preparatory discussions in progress towards it, that sustainable management of natural tropical forests in the fullest sense, as advocated in the ITTO guidelines, is not only desirable but achievable, given the provision of the necessary financial and institutional frameworks at both national and international levels.

References

IPCC. 1990. "Tropical Forestry Response Options to Global Climate Change." In: *Proc. Conf. Intergovernmental Panel on Climate Change, Sao Paulo, Brazil*. IPCC. 531pp.

Kemp, R.H. 1990. Consumer-related incentives to the sustainable management of natural tropical forests. In: *Proc. ITTO Seminar on Sustainable Development of Tropical Forests, Bali, Indonesia*. ITTO, Yokohama, Japan.

McNeely, J.A., Miller, K.R., Reid, W.V., Mittermeier, R.A. and Werner, T.B. 1990. *Conserving the World's Biological Diversity*. IUCN, WRI, CI, WWF-US, World Bank. 193pp.

Palmberg C. and Esquinas-Alcazar, J.T. 1988. "The role of United Nations agencies and other international organisations in the conservation of plant genetic resources." *Forest Ecology and Management*. 35 (1990) 171-197.

Poore, D., Burgess, P., Palmer, J., Rietbergen S. and Synott, T. 1989. *No Timber Without Trees: Sustainability in the tropical forest*. Earthscan Publications Ltd., London. 252pp.

COUNTRY STUDIES

ASIA

INDIA

Prepared by IUCN Staff[1]

1. Conclusions and Recommendations

From 1950 to 1980, India experienced a period of rapid and extensive deforestation. In the 1980s, Indian leaders began to take measures to check and reverse the disastrous ecological degradation taking place. A National Wildlife Action Plan was adopted by the government in 1983. This initiative contained guidelines for protected area expansion and rehabilitation. The National Wildlife Action Plan was followed in 1988 by a new National Forest Policy which declared the achievement of environmental stability to be the primary objective of all forest management and that the economic exploitation of forests was to be subordinate to this overriding consideration. The Government of India has undertaken several other important initiatives.

The problems in the management of both protected areas and production forests are worsening despite these positive steps. Protected area coverage is inadequate, incomplete and is increasingly threatened. Several important biogeographic zones are under-represented. Potential sites for new protected areas are under increasing pressure. The effective size of many protected areas is too small to assure long-term stability of numerous species of threatened flora and fauna and the surrounding areas are subject to increasingly intensive economic use.

Most of the existing parks and sanctuaries are subject to varying degrees of resource exploitation both legal and illegal. This resource exploitation (logging, fuelwood harvesting, grazing, agriculture, etc.) is conducted in a way that greatly decreases the forests' capacity to regenerate adequately, resulting in greatly reduced biological diversity values.

In production forests, supervision has been lax and logging standards are therefore poor. The basic requirements of the 1988 National Forest Policy regarding timber extraction have not been adhered to. Logging has been, and continues to be, an important factor in forest degradation. Fuelwood extraction has a much greater impact on forests, however, and cutting levels greatly exceed current sustainable harvest rates.

2. General Forest Management Issues

India possesses 37,847,000ha of "legally classified forests" which are "adequately stocked" (crown densities greater than 40%). "Woodlands" (crown density 10-40%), which cover 25,740,900ha, and mangroves covering 425,500ha can be added to this figure making a combined total of 64,013,400ha or 19% of national territory. According to classifications used by Champion and Seth (1968), there are 16 major forest types found in India. There are approximately 7 million ha of rain forest and 16 million ha of monsoon forests. These figures have been disputed and there is disagreement as to the extent of forests in India.

---

1   The IUCN staff is grateful to Duleep Matthai and Samar Singh who provided incisive comments on early drafts of this chapter.

Ninety-seven percent of forests in India are under public ownership. Eighty-five per cent of these forests are administered by forest departments of state governments. The remainder are owned by municipalities and village communities. A sizeable portion of the total forest estate is composed of artificial plantations. There are three main classes of forests:

a) "Reserve Forests" which are set aside for conservation, watershed protection and extraction of various forest products. These receive the highest level of protection;

b) "Protected Forests" which have management objectives that are similar to Reserve Forests but which permit controlled resource extraction by local people (though extraction is seldom confined to permissible limits); and

c) "Unclassed Forests" which cover all other publicly owned forests and which receive the lowest level of protection.

During the period from 1950 to 1980, large tracts of forests were converted to agriculture. As much as 150,000 ha/year of forest was officially allocated to non-forest use. In addition, large areas have been illegally converted to permanent agriculture and 6,800,000ha are affected by shifting agriculture. Much of the remaining forests are degraded by fuelwood cutting and cattle grazing.

The ecological and economic costs of the loss of forest cover has been great though the full extent of the damage has not been quantified. In the 1980s there has been a growing realisation, at all levels, of the consequences of such loss and an increased determination to address the problem. The government of India has established new laws and policies designed to slow deforestation and promote environmental stability. In general, on-the-ground implementation has been lagging.

The Forest [Conservation] Act of 1980 requires state governments to seek approval from the central government for conversion of forests. This permission is granted only after all other alternatives are considered and the economic gains are determined to be greater than the ecological and economic costs. According to government claims, this act has led to the reduction of legal forest conversion from upwards of 150,000 ha/year to 6500 ha/year. However, illegal encroachment of forest lands continues to be a grave problem.

In the late 1980s, the central government issued a circular to the states which prohibits the replacement of natural forests by monocultures and mandates the strict protection of environmentally sensitive areas.

The net rate of deforestation in India is uncertain as different authorities cite different statistics. FAO/UNEP (1981) projected a deforestation rate of 0.3% per year for the 1981-85 period. World Resources Institute (1991) cites a 2.3% deforestation rate for the 1975-81 period. The Government's own Forest Survey of India claims a net increase of forest cover for the 1980s. This calculation is based on studies of satellite imagery from 1981-83 and 1985-87. This increase was attributed to better protection and afforestation efforts (Singh *in litt.*, 1991).

One constructive initiative mandates the establishment of "Special Areas for Ecological Development" (SAEDs). These are, in effect, buffer zones adjacent to protected areas. Local people within SAEDs are given special benefits to compensate for the loss of their right to exploit the forest. Reports have indicated, however, that large-scale exploitation of forest resources continues in affected project zones.

All development projects are now required to file environmental impact assessments (EIAs). Negative impacts on forest resources are considered within these EIAs.

A new National Forest Policy was approved in 1988. This policy declares environmental stability to be the primary objective to which economic objectives are subordinated. Furthermore, it proposes a large-scale programme of wasteland afforestation which will increase fodder and fuelwood supplies thereby decreasing pressure on natural forests. Other interesting aspects of the National Forest Policy include the following:

1) Maintenance of one-third of national territory under forest cover. Sixty percent of hilly areas and 20% of flat areas are to be forested.

2) Corridors linking protected areas will be established or maintained in order to facilitate migration of species.

3) Remaining natural forests are to be preserved.

Despite the fact that the Government of India has fallen short of achieving the goals and objectives stated in this policy, it has made great strides in afforestation. Since 1985, when the National Wastelands Development Board was established, over 8.8 million ha of degraded areas had been afforested or reforested as of 1990 (Singh *in litt.*, 1991). At a cost of US$1.4 billion, this may be the largest afforestation programme in the world.

The conservation objectives of the National Forest Policy are complemented by the National Wildlife Action Plan which was approved by the Government of India in 1983. Its essential components include the following:

1) Establish a representative network of protected areas.

2) Improve management of protected areas and restore habitat.

3) Provide adequate protection of wildlife in multiple-use areas.

4) Rehabilitate threatened species.

5) Introduce captive breeding programmes.

6) Promote wildlife education.

7) Develop better research and monitoring capabilities.

8) Review and update national legislation and international conventions.

9) Formulate a National Conservation Strategy.

10) Collaborate with NGOs.

Some progress has been made towards achieving these objectives. Several schemes sponsored by the central government were launched in the Seventh Five Year Plan (1985-90) to create infrastructure facilities for promoting wildlife education and interpretation, captive breeding and rehabilitation of endangered species, control of poaching in areas outside national parks and sanctuaries and setting up of research facilities in important national parks including tiger reserves.

Project Tiger has been one notable conservation success. This initiative was begun in 1973 with the objective of conserving India's tiger population. Since then, 18 tiger reserves have been set up covering a total area of 2,800,000ha.

3. Extent, Status and Security of TPAs

According to the 1990 United Nations List of National Parks and Protected Areas, there are 59 national parks and 300 wildlife sanctuaries which total 13,481,148ha (4.5% of national territory).[1] This includes only areas greater than 1000ha. Although all habitat types are represented, several are inadequately covered. Existing gaps in protected area coverage include the sub-tropical zone of the outer Himalayas, the west coast of Kerala, the Eastern Ghats and the Nicobar Islands.

Many of the proposed parks and sanctuaries have not completed the necessary legal procedures and are therefore not yet legally constituted.

The wildlife sanctuaries, which cover 9,995,805ha, as well as many of the national parks, cannot be strictly considered "totally protected areas" (IUCN Categories I-V) since resource exploitation is allowed in these zones. Fifty-six percent of parks and seventy-two percent of sanctuaries have human populations within their boundaries. Forty-three percent of parks and 73% of sanctuaries grant resource extraction and utilisation rights to local people. Thirty-nine percent of parks and 73% of sanctuaries allow grazing. Illegal grazing is practiced in 67% of parks and 83% of sanctuaries. Finally, 16% of parks and 43% of sanctuaries allow timber extraction.

Because of the growing population and corresponding increase in economic activity, the protected area network specifically, and Indian forests in general, face several serious problems including the following:

1) Most protected areas have a small effective size.

2) In many cases, species population sizes are too small to maintain viable populations.

3) Plant and animal diversity is reduced in most forests.

4) Large areas are infected with weed species such as *Lantana camara*.

5) The Forest Survey of India estimated that 60% of the nations forests are not regenerating properly.

6) Domestic cattle compete with wild ungulates for forest forage.

7) Little political will exists to solve the difficult problem of illegal settlement in TPAs.

8) Inadequate integration of forest resource conservation with general land-use planning involving agriculture, grazing and fisheries.

The National Wildlife Action Plan proposes increasing coverage to 148 national parks and 503 sanctuaries totalling 15,134,200ha (5.1% of national territory).

4. Extent, Status and Security of Production Forests

Formerly, most logging was conducted by private interests but timber resource exploitation is increasingly being conducted by government-owned "Forest Development Corporations". Loggers are supposed to follow ecologically sensitive working plans. For instance, guidelines issued by the Government of India to the states prohibit logging over 1000 meters except in special circumstances and only then in very small patches. Furthermore, clearcutting is prohibited.

---

1   Estimates of the extent of parks and protected areas in India vary. See page 37 ("Basic Forest Statistics") for a summary of available data.

In practice, inadequate management and supervision have led to poor logging standards and over-exploitation. The consequence has been widespread reductions in forest quality and species diversity.

Timber exploitation is not as important an activity in India as it is in other ITTO countries. Other forms of forest exploitation have a much greater impact.

Thirteen times as much fuelwood is harvested in India as saw-logs. Current fuelwood requirements are estimated to be approximately 240 million m^3/year. This compares to a sustainable production capacity of 41 m^3/year. Clearly, great emphasis must be placed on plantation establishment in non-forest areas and improved natural forest management for firewood. Alternative fuel sources, such as natural gas and biogas, should be promoted.

Shifting agriculture is another major activity which impacts on forest resources. This is practiced widely in the north-eastern states. Shortened forest fallow cycles caused by increased population size and decreased land availability have severely stressed the resource base. At the same time, there are indications that the area affected by shifting cultivation is decreasing. An assessment carried out as a part of the Forest Survey of India using LANDSAT images from 1975-1984 indicated that shifting cultivation in the north-eastern states declined from 7,341,000ha to 6,285,400ha during that period. This was attributed to successful implementation of forest policies (Singh *in litt.*, 1991).

Similarly, tremendous increases in cattle densities have also threatened the long-term health of forests. Of India's 400 million head of cattle, 90 million are dependent on forest for grazing. The carrying capacity of these forests is estimated to be only 31 million head though even this level of exploitation could be expected to cause serious degradation. Grazed forests suffer heavy trampling, fodder lopping and susceptibility to fire and cattle-born disease. The regenerative capacity of these forests is seriously damaged.

A final major cause of forest loss has been the development of hydroelectric projects. By the 1980s several million hectares of forest had been lost through submergence and clearance for irrigated agriculture. Deforestation coupled with poor agricultural practices in the watersheds surrounding hydroelectric projects have increased siltation, decreased storage capacity and decreased life expectancy of these projects.

Basic Forest Statistics: India

Total Land Area:

297,319,000ha   (World Resources Institute, 1990. p.269)

297,319,000ha   (Collins *et al.*, 1991)

Total Forest Area:

64,013,400ha   (Collins *et al.*, 1991. p.146) ("Forest and Woodland")

121,494,000ha   (WRI, 1990. p.269) ("Forest and Woodland" 1985-87)

 64,200,000ha   (WRI, 1990. p.293) ("Extent of Forest and Woodland, 1980s: Total")

Deforestation Rate:

2.3%   (WRI, 1990. p.293) (1980s)

Production Forest Estate:

31,917,000ha   (WRI, 1990 p.293) ("Managed Closed Forests")

Totally Protected Area:

13,481,148ha   (IUCN, 1990. p.103)

6,743,000ha   (WRI, 1990. p.293) ("Protected Closed Forest")

13,170,318ha   (WRI, 1990. p.301) ("All Protected Areas")

13,178,700ha   (Collins *et al.*, 1991. p.135)

Number of Units:

359   (IUCN, 1990. p.103)

288   (WRI, 1990. p.301)

472   (Collins *et al.*, 1991. p.135)

Bibliography

Champion, H.G. and Seth, S. L. 1968. *A Revised Survey of the Forest Types of India.* Manager of Publications, New Delhi.

Collins, M.N., Sayer, J.A. and Whitmore, T.C. 1991. *The Conservation Atlas of Tropical Forests: Asia and the Pacific.* Macmillan Press Ltd., London. 256pp.

FAO/UNEP. 1981. *Tropical Forests Resources Assessment Project. Forest Resources of Tropical Asia.* Vol. 3 of 3 volumes. FAO, Rome.

IUCN. 1990. *1990 United Nations List of National Parks and Protected Areas.* IUCN Gland, Switzerland and Cambridge, UK. 275pp.

MacKinnon, J. and MacKinnon, K. 1986. *Review of the Protected Areas System in the Indo-Malayan Realm.* IUCN, Gland, Switzerland and Cambridge, UK. 284pp.

Rodgers, W.S. 1991. Protected area networks, conservation adequacy and management directions: information from India. *Tiger Paper* April-June:5-10.

Singh, S. 1991. *In litt.*, 2 December.

World Resources Institute (WRI). 1990. *World Resources: 1990-91.* Oxford University Press, Oxford. 383pp.

INDIA
TOTAL FOREST AREA

Non-Forest
233,305.6

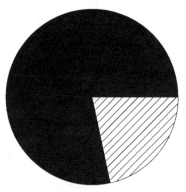

Forest
64,013.4

Note: values given in 1000s of ha

INDIA
LAND USE DESIGNATIONS

Other
283,837.9

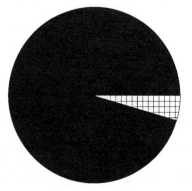

TPAs
13,481.1

Note: values given in 1000s of ha

INDONESIA

Based on the Work of
Benni H. Sormin

1. Conclusions and Recommendations

Indonesia possesses 144 million ha of legally recognised "forest land".[1] However, approximately 20 million ha of this area is not forested. Within the legally recognised forest zone, there are 18.7 million ha of totally protected areas. This network of TPAs is complemented by 30.3 million ha of Protection Forests which exist primarily to conserve watersheds. The conservation status of these areas is not satisfactory and much has to be done to protect them from conversion to alternative land uses. Already significant portions of Protection Forests and Conservation Forests have been degraded or deforested.

Similarly, the Production Forests of Indonesia, which total 60.4 million ha, have not been optimally managed. The original concept of the Government of Indonesia whereby concessionaires would act as responsible stewards of the forests has not proved effective. Shortages of Ministry of Forestry field staff and lack of motivation on the part of concessionaires are largely to blame.

The Government of Indonesia has taken positive steps to correct some of the problems in the sector.

2. Extent, Status and Security of TPAs

Indonesia is outstanding in terms of biological diversity richness. It encompasses three vegetation regions with highly diverse flora and fauna:

1) **Asia Region**: This covers Sumatra and Kalimantan and is dominated by the family Dipterocarpaceae.

2) **Australasian Region**: Includes Irian Jaya, Moluccas and the Lesser Sundas. This region is characterised by the dominance of the families Araucariaceae and Myrtaceae.

3) **Transitional Region**: Principally Sulawesi and Java. The families Myrtaceae and Verbenaceae are dominant in the transitional region.

Indonesia possesses 10% of the world's rain forest and 40 to 50% of the rain forests of Asia. These forests contain approximately 4000 tree species, 267 of which are considered commercial timber species. Indonesia's forests are also home to 500 species of mammals (including 100 endemic species) and 1500 species of birds (representing 17% of the worlds avifauna) (see Table 1).

This rich biological region is protected by a network of parks and reserves totalling 18.7 million hectares. This TPA system is complemented by 30.3 million hectares of Protection Forests whose

---

1   There is considerable disagreement regarding the various statistics on Indonesian forest resources. The numbers used in this report differ somewhat from those quoted in other sources. See pages 45-46 ("Basic Forest Statistics") for a summary of available data on Indonesian forests.

Table 1:	Biological Diversity Richness of Indonesia		

No.	Category	Total	No. of Species Protected
1.	Mammals	750	100
2.	Birds	1250	372
3.	Amphibians/Reptiles	600	28
4.	Pisces	9000	6
5.	Arthropods/Insects	12000	20
6.	Molluscus	2000	–
7.	Other invertebrates	700	–
	Sub Total	26,300	526
8.	Flowering/seed-bearing plants	25,000-30,000	36
	TOTAL	51,000-56,000	572

Source: Directorate Bina Programme PHPA (1984) and Sastrapradja (1989)

primary function is to conserve important watersheds but which also serve as reservoirs of biological diversity. The conservation status of existing reserves and watersheds is unsatisfactory and much work needs to be done to improve management and protection of these areas.

There are 36 priority watersheds located mainly in western and central parts of Indonesia. These watershed networks embrace various types of land uses. Commercial timber harvests are prohibited in Protection Forests but harvesting occurs in these areas nonetheless. Some of these areas have been converted to non-forest use.

The following criteria and considerations are used in the designation of Protection Forests:

1) Slopes greater than or equal to 45%.

2) Highly erodible soils such as regozols, lithozols and organozols with slopes greater than or equal to 15%.

3) Buffer zones along rivers and around springs at least 100 meters from the borders.

4) Any areas over 2000 meters.

5) Other special considerations deemed important by the forestry authority.

Because of these criteria, the Protection Forests are biased towards the hilly and high altitude areas encompassing various ecosystem types. Despite this bias, these areas form an important potential complement to the TPA network in terms of conservation of biological diversity.

3. Extent, Status and Security of Production Forests

Currently the annual timber production from natural forests in Indonesia is about 31 million m^3. Most of this timber (about 27 million m^3/year) comes from Production Forests. Another 4 million m^3/year comes from Conversion Forests (see definitions below). On Java the plantation forests of the state forest corporation, "Perum Perhutani", which covers about 2 million ha, produce 1.4 million m^3/year of timber. The growing stock of presently commercial timber in Indonesia has been estimated as 2000 to 3095 million m^3 (Mok 1990).

There are 64.4 million ha of permanent Production Forests representing 45% of the total forest land area. To understand how the Production Forests were identified, one must understand the government initiative of the early 1980s which set forest land use categories according to consensus. Basically, the government did not preset the targets for production forests. Instead, it applied specific criteria to determine the suitability of an area for this purpose. If the land met the criteria, then it was designated as Production Forests, otherwise, it was allocated for some other use. Determination of forest land allocation was based on a series of decrees by the Minister of Agriculture issued in 1980 and 1981. These decrees were derived from the Basic Forestry Act Number 5 (1967). The land use designations were based on province by province negotiations under the coordination of the Minister of Agriculture. Except for reserves, national parks and recreation forests, the criteria used in negotiations were as follows:

1)   Location and condition of the forest (such as timber potential).

2)   Topography/slope.

3)   Soil characteristics, particularly erodibility.

4)   Other specific considerations.

A weighted system was used to quantify the classification of the areas being evaluated. The main factors were: slope (weighted 20), soil type/erodibility (weighted 15) and rainfall intensity (weighted 10) (see Table 2). If the score of an area is 125-174, then it was designated as Limited Production Forest (see definition below). Lands with a score of below 124 were designated as Unlimited Production Forests where clearcutting is allowed.

Table 2:   The Classification of Topography, Soil Erodibility and Rainfall Intensity in Relation to Determination of Forest Land-use in Indonesia

Class	Topography	Soil Type/Erodibility	Peak Rainfall (mm/day)
1	0-8%	Alluvial, glei pluvial, gray hidromorf, laterite air tanah	0-13.6
2	8-15%	Latosol	13.6-20.5
3	15-25%	Brown forest soil, non-calcic brown, meditern	20.5-27.7
4	25-45%	Andosol, laterite, grumosol, podsol, podsolic	27.7-34.8
5	45+%	Regosol, lithosol, organosol, rezina	34.8+

Source: Decree of the Minister of Agriculture No. 837/Kpts/Um/II/1980 on the Criteria and Determination of Protection Forests, 24 November1980

By using the above considerations in the negotiations conducted in every province, the end product was the Land Use By Consensus (TGHK). The results were as follows:

1)   **Protection Forests**: (30.3 million ha) Intended primarily for watershed protection. Even though these forests are not managed administratively under the wildlife and nature conservation authority (PHPA), they function as an important complement to the TPA network and play an important role in the preservation of biological diversity.

2) **Conservation Forests**: (18.7 million ha) This includes reserves, national parks and recreation areas. Conservation Forests are intended for biological diversity conservation, scientific, cultural and recreational purposes.

3) **Limited Production Forests**: (30.5 million ha) These areas are set aside mainly for timber production, where only selective cutting is allowed.

4) **Unlimited Production Forests**: (33.9 million ha) These areas are also designated for timber production, but in Unlimited Production Forests clear cutting is allowed if necessary.

5) **Convertible Production Forests**: (30.5 million ha) Intended for timber exploitation and eventual conversion to agriculture.

According to the TGHK, there are 144 million ha of forests in Indonesia. At least 20 million of these hectares, however, are not forested. There has been considerable erosion of the forest resource base in Indonesia. For this reason, the figures noted above need to be adjusted downwards to reflect the true extent of forest cover.[1]

It is important to note that the designation of Conservation Forests was based on ecological criteria not on the weighted scoring system described above. Existing management units in this category are spread relatively evenly all over Indonesia.

Since the main considerations used in determining Production Forests are topography, soil type/erodibility, rainfall and extraction feasibility, these areas are biased to lowland forest ecosystems which are relatively high in standing timber stocks, are densely vegetated, highly diverse in flora and fauna and which are generally located in relatively flat areas with fertile soils. Therefore, it is especially critical that these areas be wisely managed so that they may complement TPA networks.

The Government of Indonesia decided that Production Forests should be developed (i.e. exploited) by private companies operating under government supervision. In 1967, the government opened up the outer islands to timber concessionaires. Currently there are 576 concessions covering over 60 million ha.

The concessionaire system gives loggers twenty-year harvesting rights over a specific area of forest land. Concession holders are legally obliged to regenerate and care for the forest. Regulations require that the processing and marketing of timber be conducted in accordance with a plan agreed upon with the Minister of Forestry. Management of the concession area is supposed to be undertaken according to three types of working plans: a twenty-year overall work plan, five-year work plans and annual work plans. These plans theoretically incorporate conservation concerns and utilise sustained yield principles.

There are two logging systems currently being used in Indonesia, namely:

1) **Clearcutting System**: This consists of clearcutting with artificial regeneration and its modified version for mangrove forests.

2) **Indonesian Selective Cutting and Planting System (TPTI)**: Designed for dipterocarp forests with normal stem diameter frequency distribution where there is sufficient natural regeneration.

---

1   For recent estimates of forest cover within the different TGHK designations, see pages 45-46 ("Basic Forest Statistics").

The clearcutting system with artificial regeneration is widely used in plantation development (pine and teak) in Java. The TPTI system, which was developed for dipterocarp forests, has been inappropriately applied to other types of forests as well.

In addition to prescribed silvicultural systems, concessionaires are expected to adhere to restrictions on areas within concessions which they are allowed to cut. For example, logging is prohibited in the following areas:[1]

1) The edges of water springs.

2) Within 50m to 200m buffer zones along river courses (according to the size of the river).

3) In buffer zones around lakes and reservoir.

4) Near the edges of steep lands to at least twice the depth of those lands.

The Director General of Forest Utilisation of the Ministry of Forestry announced another valuable measure in a letter to concessionaires, dated 12 May 1990, which required them to leave unlogged buffer forests 500m wide adjacent to marked boundaries of protected areas and 1000m buffers for unmarked boundaries.

In 1979, the Director of PPA (Nature and Wildlife Conservation) launched an initiative requiring 1-2% of concession areas to be set aside for wildlife refuges and sanctuary areas. This initiative was given added legal force by a decree of the Minister of Agriculture in 1981 which attempted to integrate biological conservation objectives into production forest by setting aside areas within concessions where logging would be prohibited. This very laudable decree was never enforced because it was no longer considered urgent. The conservation of biological diversity was felt secure in the gazetted 49 million hectares of Conservation and Protection Forests. Still, the concept has not perished. In 1989, the Director General of Forest Utilisation ordered all concessions to set aside 100ha of seed stands for each five-year plan cutting area. These stands are to be selected from the least disturbed virgin forest and must be fenced and posted.

It is not clear to what extent the regulations mandating the preservation of unlogged areas within concessions are enforced but at least one concessionaire in Pula Laut, Kalimantan, PT Inhutani, has set aside some patches of forest for biological diversity conservation and seed production.

The Minister of Forestry announced another potentially constructive decree in 1989 which made the filing of an environmental impact assessment (EIA) mandatory for all existing and proposed timber concessions and industrial tree plantations. The coordination of EIA implementation is the responsibility of the Director General of Forest Protection and Conservation (PHPA). As of August 1990, the Director General of PHPA was in the process of evaluating EIAs for 10 concessions. Forty three additional EIAs were being prepared as of that date.

In general, the concessionaire system has not been effectively implemented due to lack of staff necessary to supervise operations in the field and a lack of motivation on the part of the concessionaires.

Even if the system were implemented according to existing legal requirements, there would still be a degradation of species diversity in logged areas even though timber yields might be sustained. To comply with existing silvicultural prescriptions, some non-commercial flora

---

1   Decree of the Minister of Forestry No. 353/Kepts-II/86

species would be eliminated to provide more growing space for selected commercial species. This practice is prone to reduce biological diversity.

Other problems exist as well. For instance, relatively little research has been carried out on the effect of logging on biological diversity. Indeed, relatively little information exists on biological diversity within production areas. Only tree species which are commercially valuable are recorded in forest inventories.

The most serious threat to the permanent forest estate arises from the dynamic relation between logging activity and shifting cultivation. In most cases, the protection of logged-over areas from shifting cultivation is relatively ineffective. There is a tendency for local people to settle in areas made accessible by logging roads. It is obvious that Production Forests need more protection measures if they are to achieve biological diversity conservation objectives in the future. More systematic and close coordination with relevant external ministries is mandatory in order to achieve maximum protection of these areas in the long run.

The measures taken by the Government of Indonesia to address the problem of shifting agriculture include a campaign to raise awareness, extension programmes to encourage permanent rather than shifting agriculture, social forestry activities, involvement of local people in forest management and increased law enforcement measures.

To this end, the Government of Indonesia has intensified the extension programmes of the Directorate of Extension for Land Rehabilitation and Reforestation (RRL) and the Directorate of Extension for Natural Resource Conservation. In a similar vein, the Government of Indonesia has been conducting practical short training courses at forestry training centres for many years. The curriculum is geared towards promoting permanent agricultural practises.

In Kalimantan, an interesting small-scale pilot project has been set up called the "Concession Guided Village" ("Desa Binaan HPH"). Under this programme, concessionaires act as a "stepfather" to people who have encroached on concession lands. They teach settlers permanent agricultural techniques and involve them in plantation establishment.

Resettlement of people who encroach on production forests may sometimes be necessary. The key issue is how to make shifting cultivators appreciate the legal status of the Production Forests and to teach them that there are better methods of soil cultivation.

Relatively modest changes in the ways in which concessionaires manage forests and logging operations could contribute enormously to increasing the value of production forests for biological diversity conservation. To attain this objective the following actions are recommended:

1) Codes of practice for logging and silviculture based on the ITTO guidelines for sustainable forest management should be adopted. For this purpose an operationally-oriented set of guidelines need to be developed specifically for Indonesia.

2) The management of protected areas and production forests needs to be harmonised in an overall land-use management plan. This effort would be facilitated by the fact that both Production Forests and Conservation Forests are under the jurisdiction of the Ministry of Forestry.

3) Development of multiple-use forests based on good examples found in Java.

4) Regulations requiring environmental impact assessments and the establishment of buffer forests need to be enforced.

5) Efforts in research, extension and training on biological diversity conservation need to be intensified.

Proposed activities for ITTO include the following:

1) Support for field and research activities to improve management of concessions especially with regard to improving national capacity to conduct biological inventories and the integration of biological diversity conservation objectives in production forests.

2) Assist in the operationalisation of the ITTO guidelines for forest management. A series of field manuals need to be developed for each region.

3) Assist in the monitoring and establishment of protected areas and multiple-use forests. Monitoring could be conducted by the Forest Conservation Programme of IUCN in conjunction with the World Conservation Monitoring Centre.

4) Provide consulting services to governments and concessionaires to improve their management capability.

5) Help to reduce the volume and increase the value of timber products from tropical forests.

6) Publish a directory of production areas which meet minimum conservation standards.

Basic Forest Statistics: Indonesia

Total Land Area:

181,157,000ha (World Resources Institute, 1990. p.269)
181,157,000ha (Collins *et al.*, 1991. p.146)
193,027,000ha (Burgess, 1988. p.6)

Total Forest Area:

144,000,000ha (Sormin, 1990. p.1) (20,000,000 of the total officially classified forest area is not forested or "unproductive")
18,020,000 ha (WRI, 1990. p.269) ("Forest and Woodland" 1985-87)
116,895,000ha (WRI, 1990. p.293) ("Extent of Forest and Woodland, 1980s: Total")
117,914,000ha (Collins *et al.*, 1991. p.146)
143,970,600ha (Burgess, 1988. p.9)
108,600,000ha (MoF/FAO, 1990b.)
117,928,300ha (RePPProT, 1990.) (This estimate excludes the sparsely forested islands of Java, Bali and the Lesser Sundas)
120,671,300ha (Dillenbeck, 1991) (This estimate combines data from the RePPProt study and Collins *et al.*)

Deforestation Rate:

0.5% (WRI, 1990. p.293) (1980s)
0.5% (derived from: Collins *et al.*, 1991.) ("1981-85")
1.0% (Collins *et al.*, 1991. p.141) ("late 1980s")

Production Forest Estate:

64,400,000ha (Sormin, 1990. p.1)
40,000ha (WRI, 1990. p.293) ("Managed Closed Forests")
64,391,000ha (Collins *et al.*, 1991. p.143)
64,403,600ha (Burgess, 1988. p.6)
46,100,000ha (MoF/FAO, 1990b) (This is an estimate of the extent of legally recognised Production Forest which remains forested)
49,324,500ha (RePPProT, 1990.) (This is an estimate of the extent of legally recognised Production Forest which remains forested. This estimate excludes the sparsely forested islands of Java, Bali and the Lesser Sundas)

Watershed Protection Forests:

30,300,000ha	(Sormin, 1990. p.1)
30,316,000ha	(Collins *et al.*, 1991. p.143)
25,200,000ha	(Mof/FAO, 1990b) (This is an estimate of the extent of legally recognised Protection Forest which remains forested)
25,191,000ha	(RePPProT, 1990) (This is an estimate of the extent of legally recognised Protection Forest which remains forested)

Totally Protected Area:

18,700,000ha	(Sormin, 1990. p.1)
5,430,000ha	(WRI, 1990. p.293) ("Protected Closed Forest: 1980s")
14,067,051ha	(WRI, 1990. p.301) ("All Protected Areas")
17,799,787ha	(IUCN, 1990. p.110)
14,600,000ha	(MoF/FAO, 1990b.) (This is an estimate of the extent of legally recognised Conservation Forest which remains forested)
17,317,300ha	(RePPProT, 1990.) (This is an estimate of the extent of legally recognised Conservation Forest which remains forested)
17,500.000ha	(Collins *et al.*, 1991. p.155)
19,806,000ha	(Collins *et al.*, 1991. p.163)

Number of Units:

141	(WRI, 1990. p.301)
169	(IUCN, 1990. p.110)
320	(Collins *et al.*, 1991. p.155)

Bibliography

Burgess, P.F. 1988. Natural Forest Management for Sustainable Timber Production: The Asia/Pacific Region. Unpublished report prepared for IIED and ITTO.

Collins, M.N., Sayer, J.A. and Whitmore T.C. 1991. *The Conservation Atlas of Tropical Forests: Asia and the Pacific*. Macmillan Press Ltd., London. 256pp.

Department Kehutanan. 1989. Pengembangan zone (Mintakat) penyangga dan daerah penyanggah kawasan konservasi. Unpublished paper presented at a working session of the Ministry of Forestry R.I.

Dillenbeck, M.R. 1991. Forest management in Indonesia with special emphasis on production forests and parks and protected areas. IUCN-US. Unpublished report.

Director, Perlindungan dan Pengawetan Alam, Ditjen Kehutanan. September, 1979. Peranan kawasan konservasi dalam bidang pelestarian hutan Bogor. Unpublished paper presented at special meeting with concession holders.

Directorate Inventarisasi Hutan. 1990. Tegakan hutan Indonesia. Jakarta, proyek inventarisasi dan evaluasi sumber daya alam. Unpublished report.

Government of Indonesia and IIED 1985. A review of policies affecting the suitable development of forest land in Indonesia. IIED. Unpublished report.

IUCN. 1990. *1990 United Nations List of National Parks and Protected Areas*. IUCN, Gland, Switzerland and Cambridge, UK.

Ministry of Forestry and FAO. 1990a. *Indonesia National Forestry Action Plan* (country brief). Ministry of Forestry, Jakarta.

Ministry of Forestry, Government of Indonesia and the Food and Agriculture Organization of the United Nations (MoF/FAO). 1990b. *Situation and Outlook of the Forestry Sector in Indonesia, Volumes 1-4*. Ministry of Forestry, Jakarta.

Ministry of Forestry R.I. 1990c. *Indonesian Tropical Rain Forest Conservation Areas*. Ministry of Forestry, Jakarta.

Ministry of Forestry. 1979. *Forestry Indonesia: Jakarta Development of Planning and Forestry Information System Project*.

Mok, Sian Tuan. 1990. Sustainable management and development of tropical forest in ASEAN. Unpublished paper presented at the ASEAN Seminar on Management of Tropical Forests for Sustainable Development.

Oliva, R.V. 1987. Harmonisation of forest policies in the ASEAN region. ASEAN Institute of Forest Management, Kuala Lumpur. Unpublished paper.

RePPProT (Regional Physical Planning Programme for Transmigration). 1990. *The Land Resources of Indonesia: A National Overview*. Publisher not identified.

Sormin, B.H. 1990. The role of production forests in biological diversity conservation in Indonesia. Unpublished report prepared for the IUCN Forest Conservation Programme workshop on "Realistic Strategies for Tropical Forest Conservation" in Perth, Australia.

World Resources Institute (WRI). 1990. *World Resources: 1990-91*. Oxford University Press, Oxford. 383 pp.

UNDP/FAO 1981. *National Conservation Plan for Indonesia*. Bogor Field Report 19.

INDONESIA
TOTAL FOREST AREA

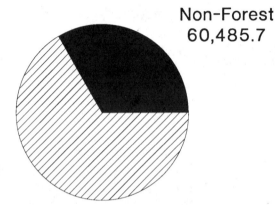

Non-Forest
60,485.7

Forest
120,671.3

Note: values given in 1000s of ha

INDONESIA
LAND USE DESIGNATIONS

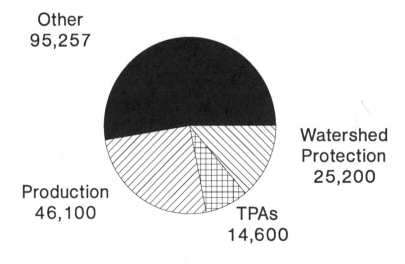

Other
95,257

Watershed
Protection
25,200

Production
46,100

TPAs
14,600

Note: values given in 1000s of ha

MALAYSIA

Based on the work of Thang Kooi Chiew

1. Conclusions and Recommendations

Malaysia has a very extensive permanent forest estate complemented by an extensive and growing system of TPAs. The major constraints to biological diversity conservation are that harvesting is occurring at above the sustainable yield and premature re-entry into logged-over forest and over-harvesting are frequent. Logging practices are also destructive, mainly because of the high density of commercial timber and the hilly nature of the terrain. Replacement of natural forests with plantation of exotic species in Sabah and with oil and rubber in the Peninsula are eroding the natural forest estate.

The main priorities for biodiversity conservation are expansion and consolidation of the TPA system to include all forest ecosystems and stricter application of logging regulations in the permanent forest estate (PFE). Revival and extension of the Virgin Jungle Reserve network would make a major contribution by creating refuges for wildlife in logged-over forests.

2. Extent, Status and Security of TPAs

One national park and 21 equivalent reserves cover 740,000ha in Peninsular Malaysia (6% of the land area).[1] In Sarawak seven national parks and three wildlife sanctuaries cover 260,000ha and in Sabah six national parks, and seven equivalent reserves cover 390,000ha (3% and 5.6% respectively). Malaysia as a whole thus has a total of 1,390,000ha of TPAs. The standard of management of these areas is variable. The wildlife reserves and sanctuaries in the Peninsula have suffered some illegal logging. Small areas have been excised from the Kinabalu National Park in Sabah. Major extensions to the protected area system are planned for Sarawak (a further 606,960ha bringing the state coverage to 8.33%) but successful protection will depend upon the resolution of numerous claims for native customary rights. Logging concessions already cover significant parts of the area. A number of new priority areas have been identified in Sabah and a State Conservation Strategy now being developed will promote their protection. The highest priority in the Peninsula is to achieve adequate protection of the existing wildlife reserves and sanctuaries (existing legislation only protects species, not habitat) and to secure protection for a small number of new sites to complete representative coverage of all forest types. Mangroves, swamp forests and lowland dipterocarp forests are at present under-represented in the system. If adequate management standards could be achieved for existing TPAs and those priority areas already identified could be added to the system, Malaysia would have a satisfactory network of TPAs.

The PFE is very extensive and plays a major role in conserving biological diversity in the federation. It covers most categories of forest land and forest reserves are often contiguous with

---

1   IUCN has been unable to substantiate the forest data used in this report which is inconsistent with that given by other authors such as Collins *et al*. A summary of the forest statistics available from various sources is provided at the end of this chapter.

TPAs and thus enhance their conservation value by forming buffer zones. Most of the PFE is under natural forest, much of which has been logged. There have been recent moves to greatly expand plantation forestry, especially in Sabah, and monocultures of *Acacia mangium, Albizzia falcataria, Eucalyptus* spp. etc. now cover significant areas of the PFE thus reducing its value for biodiversity. Shifting cultivation occurs on 120,000ha in the PFE in Sarawak and 5000ha of primary forest within the PFE are lost each year to this cause. Shifting agriculture is a relatively minor problem in Sabah and has little impact in the Peninsular. Significant areas of the PFE in the Peninsular, particularly in the valuable lowland forests, have been alienated for rubber and oil palm plantations.

Sarawak plans to increase its PFE to cover 70% of the state territory and to manage the entire area under a selective harvesting system.

3. Extent, Status and Security of Production Forests

Malaysian forests have a high volume of tree species, mainly dipterocarps, which are much sought after for industrial timber production. Timber yields average $38m^3$/ha in Sarawak and up to $90m^3$/ha in Sabah for the first cut. Informal estimates suggest that, at present, timber production is as much as four times the level which can be sustainably produced from the PFE. Much of the timber comes from forests that are scheduled for conversion to agriculture and this supply will soon be exhausted. Timber exports are the mainstay of the economies of Sabah and Sarawak and the economic problems that will come with the inevitable decline of this industry in the relatively near future will have serious impacts on those states. There is a danger that this will create pressures to over-harvest the PFE or even to log parts of the TPAs.

The high volume of timber extracted from Malaysian forests means that disruption at the time of logging is inevitably quite severe. This is especially so in Sabah where most of the large trees are removed during the first cut. In Sarawak forests, the age distribution of the trees favours more selective systems of management and damage during logging is said to be less. Forests in Sarawak and those in Sabah managed under polycyclic systems, are much more susceptible to re-logging before completion of the prescribed logging cycle and work in Sabah has shown this to be particularly harmful to biodiversity values.

Silvicultural treatments have been practiced over large areas of Malaysia's forests but their impacts on biodiversity have not been measured. Forests managed intensively under the Malaysian Uniform System tend towards almost pure stands of even-aged dipterocarps and will presumably support a less diverse fauna and flora than forests managed under the selection systems. In the latter system, some liberation thinning was practiced in the past and presumably this would have eliminated some important food species for fauna (*Ficus* spp. for example). There is some evidence that biodiversity declines after the second and successive logging cycles under both systems because there is insufficient time for canopy epiphytes to recolonise the forest between cycles.

A system of 119 virgin Jungle Reserves cover 109,571ha in Peninsular Malaysia and Sabah. These are left unlogged as sites for research and for the conservation of genetic resources. They could make a valuable contribution to biodiversity conservation as sources of colonists for adjacent logged-over forests. Their present distribution is not optimal for this function (some are isolated from the main forest blocks) and many have suffered disturbance from illegal logging.

Present measures to apply forest management regulations more stringently will greatly enhance the biodiversity potential of the PFE and Malaysia has the potential to be a model of optimal allocation of forest land for this purpose.

Table 1: **Distribution and Extent of Major Forest Types in Malaysia, 1989 (million ha)**

Region	Land Area	Dipterocarp	Swamp	Mangrove	Total Forested Land	Percentage Total of Forested Land
Peninsular Malaysia	13.16	4.96	0.46	0.11	5.51	41.9
Sabah	7.37	3.93	0.19	0.32	4.44	60.2
Sarawak	12.33	7.04	1.24	0.17	8.45	68.5
Malaysia	32.86	15.91	1.89	0.60	18.40	56.0

Table 2: **Permanent Forest Estate in Malaysia, 1989 (million ha)**

Region	Protection Forest		Production Forest		Total Land Area under PFE		Total Land Area	Percentage of Total Land Area
	Virgin	Logged	Virgin	Logged	Virgin	Logged		
Peninsular Malaysia	1.90	–	0.44	2.40	2.34	2.40	4.74	36.0
Sabah	0.35	–	0.75	2.25	1.10	2.25	3.35	45.5
Sarawak	0.49	–	2.71	1.44	3.20	1.44	4.64	37.6
Malaysia	2.74	–	3.90	6.09	6.64	6.09	12.73	38.7

Basic Forest Statistics: Malaysia

Total Land Area:

32,860,000ha	(Thang Kooi Cheiw, 1990)
32,855,000ha	(World Resources Institute, 1990. p.269)
13,159,800ha	Peninsular Malaysia (Collins *et al*. 1991. p.183)
19,870,000ha	Sabah and Sarawak (Collins *et al*. 1991. p.201)

Total Forest Area:

5,510,000ha	Peninsular Malaysia (Thang Kooi Chiew, 1990)
12,890,000ha	Sabah and Sarawak (Thang Kooi Chiew, 1990)
15,178,000ha	(WRI, 1990. p.269) ("Forest and Woodland: 1985-87")
20,966,000ha	(WRI, 1990. p.293) ("Extent of Forest and Woodland, 1980s")
6,975,000ha	Peninsular Malaysia (Collins *et al*. 1991. p.183) ("Rain Forest")
13,067,000ha	Sabah and Sarawak (Collins *et al*. 1991. p.201) ("Rain Forest")
6,353,200ha	Peninsular Malaysia (Burgess, 1988. p.6)
13,925,300ha	Sabah and Sarawak (Burgess, 1888. p.6)

Deforestation Rate:

1.2%	(WRI, 1990. p.293) (1980s)
1.1%	(derived from: Collins *et al*. 1991. p.183.)

Production Forest Estate:

2,840,000ha	Peninsular Malaysia ("Production Forests") (Thang Kooi Chiew, 1990)
7,150,000ha	Sabah and Sarawak ("Production Forests") (Thang Kooi Chiew, 1990)
2,499,000ha	(WRI, 1990. p.293) ("Managed Closed Forests:1980s")
2,850,000ha	Peninsular Malaysia (Collins *et al*. 1991. p.185) ("production forest" 1985)
3,450,700ha	Peninsular Malaysia (Burgess, 1988. p.6)
6,238,400ha	Sabah and Sarawak (Burgess, 1988. p.6)

Watershed Protection Forests:

1,900,000ha	Peninsular Malaysia (Thang Kooi Chiew, 1990)
840,000ha	Sabah and Sarawak (Thang Kooi Chiew, 1990)
1,900,000ha	Peninsular Malaysia (Collins *et al*. 1991. p.185) ("protection forest" 1985)
1,067,900ha	Peninsular Malaysia (Burgess, 1988. p.6)
3,020,000ha	Sabah and Sarawak (Burgess, 1988. p.6)

Totally Protected Area:

740,000ha	Peninsular Malaysia (Thang Kooi Chiew, 1990)
650,000ha	Sabah and Sarawak (Thang Kooi Chiew, 1990)
959,000ha	(WRI, 1990. p.293) ("Protected Closed Forest: 1980s")
1,101,353ha	(WRI, 1990. p.301) ("All Protected Areas")
1,162,204ha	(IUCN, 1990. p.127)
829,300ha	Peninsular Malaysia (Collins *et al*. 1991. p.188)
568,600ha	Sabah and Sarawak (Collins *et al*. 1991. p.208)

Number of Units:

44	(Thang Kooi Chiew, 1990.)
39	(WRI, 1990. p.301)
45	(IUCN, 1990. p.127)

Bibliography

Burgess, P.F. 1988. Natural forest management for sustainable timber production: the Asia/Pacific region. Unpublished report prepared for IIED and ITTO.

Collins, N.M., Sayer, J.A., and Whitmore, T.C. 1991. *The Conservation Atlas of Tropical Forests: Asia and the Pacific*. Macmillan Press Ltd., London. 256pp.

IUCN. 1990. *1990 United Nations List of National Parks and Protected Areas*. IUCN, Gland, Switzerland and Cambridge, UK.

Thang Kooi Chiew. 1990. Untitled. Unpublished report prepared for the IUCN Forest Conservation Programme workshop, "Realistic Strategies for Tropical Forest Conservation" in Perth, Australia.

Whitmore, T.C. and Sayer, J.A. (in press). Deforestation and species extinction in tropical moist forests. In: *Tropical Deforestation and Species Extinction*. Proceedings of IUCN General Assembly workshop. IUCN, Gland, Switzerland.

World Commission on Environment and Development (WCED). 1987. *Our Common Future*. Oxford University Press, Oxford, U.K.

World Resources Institute (WRI). 1990. *World Resources: 1990-91*. Oxford University Press. Oxford. 383pp.

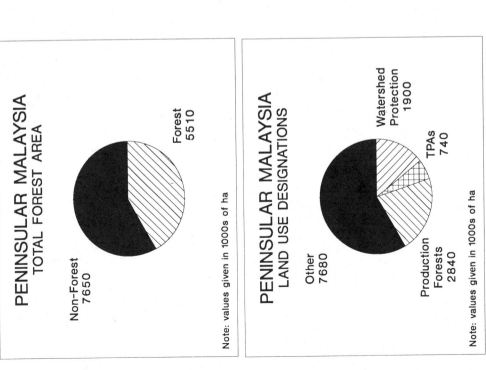

NEPAL

Based on the work of Rabbi Bista

1. Conclusions and Recommendations

Situated at the intersection of the Palearctic and the Indo-Malayan biogeographical realms and with a wide altitudinal range, Nepal hosts a great variety of ecosystems. Its high human population density (ca. 120/km^2), however, exerts strong pressures on the natural environment and threatens to impoverish its rich biodiversity. In spite of scarcity of land for the rapidly growing populace, the country can boast an outstanding record of conservation: at present 7.4% of the national territory is legally secured in National Parks (7) and Reserves (5). With an additional two multiple-use conservation areas expected to be gazetted soon, the TPA system will cover approximately 1,660,000ha, or 11.3%, of the country. Fifteen per cent of the natural forests in the Terai and Siwaliks – the two regions which contain tropical and subtropical forests – are currently protected in two National Parks and three Wildlife Reserves. Management is seriously hampered by difficult access and insufficient technical staff and logistical support. Because the dense rural population is heavily dependent on access to natural resources, preventing encroachment and illegal use of TPAs has been a main objective of TPA management. Opposing local interests have escalated, and a main objective of future management is to reconcile conservation interests with those of local communities. Notable attempts have been initiated, including developing management plans for surrounding buffer zones.

An estimated 5.5 million ha, or 37% of the national territory, is currently under natural forest cover, although a large share of this is in a highly degraded condition. Deforestation has been pronounced: in the two physiographic zones containing tropical and subtropical forests, approximately 7.5% of the forests were cleared during a seven year period 1978-1985. No areas have yet been set aside for permanent forestry production. A newly adopted Master Plan for the Forestry Sector (1988) aims at halting deforestation and increasing production for local consumption, mainly through strengthening the present community-based forestry programme. Besides plantation forestry and enrichment of natural forests, this plan also includes provisions for limited national and "leasehold" forestry enterprises, mainly targeted for community demonstration, soil and watershed protection and for supplying products for forest-based industries. Because the main thrust of the forestry programme is aimed at multiple-use forestry, conservation of biodiversity is expected to receive increased attention in the management of forests outside TPAs.

2. Extent, Status and Security of TPAs

Due to altitudinal zonation, only the two lowland regions (the Terai bordering India, and the adjacent Siwaliks), contain tropical and subtropical (monsoonal) mixed hardwood forests. Together, these two zones contain about one-third of the natural forests of the country, of which 28,740ha (15%) are protected in two National Parks (Chitwan and Bardia) and three Wildlife Reserves (Sukla Phanta, Parsa and Koshi Tappu). The Koshi Tappu Reserve is situated in the Bengalian Rain Forest biogeographical province, whereas all other lowland TPAs belong to the Indo-Ganges Monsoon Forest province of Udvardy's (1975) classification. In the densely populated Midland region, the mixed hardwood and conifer forests are highly degraded, and only 3.1% of the total forest cover is protected in the form of portions of one smaller National Park

(Khaptad), a Hunting Reserve (Dhorpatan) and a watershed Conservation Reserve (Shivapuri). In the Himalayas, four National Parks (Sagarmatha, Langtang, Rara and Shey-Phoksumdo) and the Dhorpatan Hunting Reserve provide protection of the upland conifer-dominated forests. In addition, two large multiple-use Conservation Areas (Annapurna and Barun-Makalu), totalling nearly 600,000ha, are being considered for legal gazettement, which will provide added protection of the natural forests in the uplands.

Current TPAs are legally secure and they contribute significantly to foreign exchange earnings. However, the dense rural population, with a traditional subsistence lifestyle and dependence on forest produce, represents a formidable challenge to management. To prevent encroachment and other activities which are considered incompatible with traditional conservation objectives, laws which restrict local peoples access have been strictly enforced, including use of military personnel. This has fostered alienation and hostile attitudes towards park authorities and conservation efforts in general. To ease tension and develop a more harmonious relationship with the local communities, controlled, low-level utilisation of key products (thatch grass and firewood) is now permitted on an experimental basis in the two main parks in the Terai. Developing management plans for buffer zones around parks and reserves and ensuring local people's involvement and sharing of benefits, are considered high priority areas in the future management strategy of TPAs in the country. Extremely difficult access and mode of communication, coupled with inadequate resources for technical staff and logistic support, are a major constraint which must be overcome for improving management operations and TPA/local peoples relationships.

International agencies, particularly FAO, the Frankfurt Zoological Society, WWF and the Smithsonian Institution, and the recently established (1984) King Mahendra Trust for Nature Conservation have provided generous assistance to the present TPA network, which is administered and managed by the Department of National Parks and Wildlife Conservation within the Ministry of Forests and Soil Conservation.

3. Extent, Status and Security of Production Forests

According to the recently adopted Master Plan for the Forestry Sector (HMG/Nepal, 1988) the estimated 5,518,000ha of forest cover (37% of national territory) is highly degraded: one quarter has less than 40% crown closure, and two-thirds consists predominately of small-sized pole timber. The culmination of a governmental resettlement scheme, overcutting and encroachment due to land shortage has reduced the forest cover, particularly in the Midland and the Terai regions. During a seven year period (1978-1985), nearly 2 million ha (7.5% of the forests in those two zones) were cleared for agriculture and other uses. Forest degradation is not uniform; due to difficult access, particularly in the Siwaliks, degradation and encroachment are most severe in the vicinity of settlements and along the East-West Highway.

A main thrust of the new Master Plan is to halt deforestation and to develop management schemes for sustainable community forestry. Hence, the main part of the productive forest land will be managed for multiple-use purposes. This includes rehabilitation of degraded land by forest plantations, using mainly local species (*Dalbergia sissoo*, *Acacia catechu* and *Pinus roxburghii*). Community reforestation programmes have been in operation for several years, and as of 1986, approximately 69,000ha of plantations had been established.

Besides its main emphasis on community forestry, the new Master Plan also calls for a National and Leasehold Forestry Programme, as it is recognised that community forestry cannot by itself provide the increased production that is needed for commercial and industrial uses. Under "national" forestry, management will be directed by the Forestry Department, while under "leasehold" forestry some government land will be leased out for private forestry enterprise. In the Terai this will take the form of industrial forestry based on plantations and intensive management of the natural sal (*Shorea robusta*) forests. Some exotics (mainly *Albizia* spp, *Tectona grandis*, and *Eucalyptus* spp.) will be used in addition to suitable local species (khair and sissoo). In the middle hills, the programme aims at promoting social forestry through demonstration areas for enhanced production of fuelwood, timber for local construction materials, fodder and other non-timber products. A large share of the "national" forestry operations is targeted for the fragile and unstable Siwaliks and the upland, more sparsely populated parts of the country, where soil and watershed protection and ecosystem conservation will have highest priority.

Because protection of the environment has been recognised as a fundamental national objective, the Master Plan proposed to create a new National Authority for the Protection of the Environment (NAPE). Undoubtedly, this high-level governmental body will provide added assurance and incentives that Nepal's forests will be managed sustainably with due regard for their biological diversity in the future. International financing is currently being sought for the implementation of the Master Plan, and encouraging support has already been received from bilateral and multilateral donors.

Table 3. FOREST COVER AND PROTECTED AREAS (in 000 ha) IN DIFFERENT PHYSIOGRAPHIC ZONES IN NEPAL

Meters a.s.l.	Terai 60-330	Siwaliks 120-2000	Midlands 200-3000	Mount. 1-4000	Alpine >4000	Total
Total Land Area	2110	1886	4442	2960	3350	14748
% of country	14.3	12.7	30.1	20.0	22.7	(99.8)
Total Forested*	475	1438	1811	1639	155	5518
	+ 30	29	404	176	67	706
	505	1467	2215	1815	222	6224
% Forested**	22	76	41	55	5	37
	23	77	50	61	7	42
Protected Areas						
– Sukla Phanta WR	15.5					15.5
– Bardia NP	36.7	60.1				96.8
– Parsa WR	20.4	29.5				49.9
– Koshi Tappu WR	17.5					17.5
– Chitwan NP		93.2				93.2
– Khaptad NP			2.5	20.0		22.5
– Shivapuri CR			14.5			14.5
– Dhortpatan HR			51.7	70.8	10.0	132.5
– Langtan NP				61.1	109.9	171.0
– Rara NP				10.6		10.6
– Sagarmatha NP				0.7	114.1	114.8
– Shey Phoksumdo NP				7.6	347.9	355.5
	90.1	182.8	68.7	170.8	581.9	1094.3
% of total land	4.2	9.6	1.5	5.8	17.4	7.4
% of forest***	17.8	12.5	3.1	–	–	5.5
– Annapurna CA						266.0
– Barun-Makalu CA						300.0
Total TPA						1660.3
% of country						11.3

* Acreage of natural forests + degraded shrublands in zone
** Percent forest and percent forest + degraded shrublands in zone
*** Percent TPA of total forested area in zone. Assumes all TPA is forested land
  (A small proportion – 69,000ha – of plantations is included in forested areas in Terai, Siwaliks, Midlands and Mountains)

Source: Master Plan for Forestry Sector Nepal, Ministry of Forests and Soil Conservation, HMG/N, 1988

Basic Forest Statistics: Nepal

Total Land Area:

13,680,000ha   (World Resources Institute, 1990. p.269)

Total Forest Area:

5,518,000ha   (Bista, 1990)
2,308,000ha   (WRI, 1990. p.269) ("Forest and Woodland: 1985-87")
2,121,000ha   (WRI, 1990. p.293) ("Extent of Forest and Woodland, 1980s: Total")

Deforestation Rate:

4.0%   (WRI, 1990. p.293) (1980s)

Totally Protected Area:

1,012,000ha   (derived from: Bista, 1990)
330,000ha   (WRI, 1990. p.293) ("Protected Closed Forest: 1980s")
958,500ha   (WRI, 1990. p.301) ("All Protected Areas")
958,500ha   (IUCN, 1990. p.133)

Number of Units:

11   (WRI, 1990. p.300)
11   (IUCN, 1990. p.133)

Bibliography

Bista, R.B. 1990. Forestry sector development: its role in the conservation of biological diversity in Nepal. Unpublished paper prepared for the IUCN Forest Conservation Programme workshop, "Realistic Strategies for Tropical Forest Conservation" in Perth, Australia.

Food and Agriculture Organization of the United Nations, 1989. *Interpretation of the International Undertaking of Plant Genetic Resources.*

HMGN. Master Plan for the Forestry Sector, 1989. Country background.

HMGN. Master Plan for the Forestry Sector, 1989. Forestry sector policy.

HMGN. Master Plan for the Forestry Sector, 1989. Main report.

HMGN. Master Plan for the Forestry Sector, 1989. Forest development plan for the supply of main forest products.

HMGN. Master Plan for the Forestry Sector, 1989. Plan for the conservation of ecosystems and genetic resources.

HMGN. Master Plan for the Forestry Sector, 1989. Soil conservation and watershed management development plan.

HMGN. Project for the Implementation of the Master Plan for the Forestry Sector, 1990. Environmental and conservation aspects of the master plan for the forestry sector of Nepal.

HMGN. Project for the Implementation of the Master Plan for the Forestry Sector, 1990. Implementation schedule of the master plan for the forestry sector.

HMGN. Project for the Implementation of the Master Plan for the Forestry Sector, 1990. Terminal report.

IUCN. 1988. *National Conservation Strategy for Nepal.* IUCN, Gland, Switzerland.

IUCN. 1988. Draft articles prepared by IUCN for inclusion in a proposed convention on the conservation of biological diversity and for establishing of a fund for that purpose.

IUCN.1990. *United Nations List of National Parks and Protected Areas.* IUCN, Gland, Switzerland and Cambridge, UK.

King Mahendra Trust for Nature Conservation. Strategy for environmental conservation in Nepal, the initial five years (1988/89-1992/93), Action Plan of KMTNC.

UNEP. 1990. Report of the second session of the ad hoc working group of experts on biological diversity, Geneva.

UNEP. 1990. Report of the third session of the ad hoc working group of experts on biological diversity, Geneva.

UNEP. 1990. Report of the ad hoc working group during its third session in preparation of a legal instrument on biological diversity of the planet.

UNEP. 1990. Third session of the ad hoc working group of experts on biological diversity, a note by the executive director.

UNEP. 1990. Third session of the ad hoc working group of experts on biological diversity. Biological diversity: global conservation needs and costs.

UNEP. 1990. Third session of the ad hoc working group of experts on biological diversity. Current multilateral, bilateral, and national financial support for biological diversity conservation.

World Resources Institute (WRI). 1990. *World Resources: 1990-91*. Oxford University Press. Oxford, UK.

NEPAL
TOTAL FOREST AREA

Non-Forest
8162

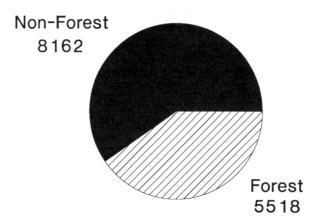

Forest
5518

Note: values given in 1000s of ha

NEPAL
LAND USE DESIGNATIONS

Other
12,721.5

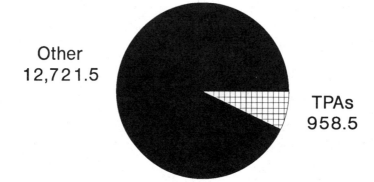

TPAs
958.5

Note: values given in 1000s of ha

PAPUA NEW GUINEA

Based on the work of S. M. Saulei

1. Conclusions and Recommendations

Due to its still largely intact forest cover and wide altitudinal variation, Papua New Guinea is exceedingly rich in biological diversity. The current TPA network, however, is small and covers only an estimated 2% of the total national territory. Most of this is contained within the Wildlife Management Areas, where certain low-level utilisation of the natural resources is practised. Less than 5000ha (0.01%) is currently under full protection in the form of National and Provincial Parks.

Due to the system of land tenure where nearly all forest land belongs to customary landowners, no permanent forest estates have yet been set aside for production forestry. Of a total forest acreage of approximately 35,200,000ha (76% of total national territory), 3,200,000ha, or 9.1%, have been selectively logged and a further 1.5 million ha is being considered for future development. Up until recently, private timber companies have normally purchased logging rights directly and/or indirectly from the landowners and have paid other fees (export licenses, etc.) to the government. A recent inquiry into the official forestry sector uncovered gross mismanagement. A newly adopted National Forest Action Plan (1991), based on a World Bank TFAP review mission, calls for major policy changes and will, if implemented, provide a framework for improved management and conservation of the forest resources in the country. Major shortcomings at present are lack of basic inventory data (forest types and associated biodiversity), insufficient knowledge of alternative silvicultural treatments and forest development following logging. Realisation of the stated objectives in the TFAP will require active involvement of the landowners at all stages of management, besides substantial international financial assistance.

2. Extent, Status and Security of TPAs

Of the country's 46.2 million ha total land area, it is estimated that about 2% or 924,000ha, have been designated for protection. These are mainly made up of National Parks (4) and Provincial Parks (4), Nature Reserves (1) and Wildlife Management Areas (11). Other management categories include smaller sanctuaries, village reserves and cultural and historical sites. The two main parks are Varirata (1063ha) and McAdam (2080ha), both on the mainland. Although of insufficient total coverage, the TPA network is fairly well distributed, ranging from the interior highlands to the outlying smaller islands. One more Provincial Park and four more Wildlife Management Areas may be gazetted in 1991, raising the TPA coverage to approximately 1.3 million ha, or 2.8% of total land area.

A major obstacle for establishing National Parks and Reserves is the intricate land tenure system in PNG, where about 95% of the total land is owned by customary land owners (through clans), the majority of whom live a subsistence or semi-subsistence lifestyle. Following a South Pacific Conference on National Parks and Reserves (IUCN, 1975), a strategy was adopted whereby protected areas should be held under customary tenure.

Subsequently, eleven Wildlife Management Areas were established, and several others have been proposed. These are by far the most important category of TPAs, together accounting for more than 90% of all protected areas in the country. Because rules and guidelines for these areas have been designed by local landowner committees to protect access to traditional resources, some low-level utilisation takes place (viz. controlled subsistence and licensed hunting, collection of megapode eggs, ornamental, medicinal plants, etc.).

Inadequate resources for staff, both at the central (Department of Environment and Conservation) and provincial level, seriously hamper effective management of the TPAs. Infrastructure for tourism is also inadequate. Only the Varirate National Park has suitable recreational facilities for visitors. All government TPAs suffer from encroachment and illegal activities, mainly arising from unsettled compensation claims and disputes between the central government and local landowners. Although the land tenure system of the country may be regarded as the main hindrance to forest development, traditional land uses based on customary land ownership may have also effectively safeguarded the forest resources from over-exploitation and alienation.

3. Extent, Status, and Security of Production Forests

Because most of the land and forest resources are owned by customary landowners, no forests have been designated as permanent production forests. Based on recent FAO calculations (1985), the total forest cover in Papua New Guinea is estimated at 36,200,000ha, or 76% of total national territory. Of this, 42%, or 15 million ha, have been considered as accessible forests for development, while 58% or 21 million ha, is inaccessible. However, 3.2 million (or 21%) of the operable forests have been logged during the last 4-5 decades, and a further 1.5 million ha are being considered for future development. The remaining categories are inoperable lowland and montane forests 10.8 million ha), mangroves (4.5 million ha), swamp woodlands (3.3 million ha) and regrowth on abandoned cultivated land and other disturbed areas (2.4 million ha), including 55,000ha of forest plantations.

While protected areas are mainly administered and managed by the Department of Environment and Conservation (in the case of Wildlife Management Areas by local landowners' committees), commercial forestry is administered by the Department of Forests. Landowners sell logging rights to private timber operators as well as to the Department of Forests, which later sells the rights to the developers through a concession system administered by this Department. Ongoing practices, particularly those relating to industry and trade, have been under public criticism for several years. An official inquiry (Barnett Commission, 1989), and a World Bank Tropical Forestry Action Plan mission (1990) both concluded that drastic changes were needed to ensure social, economic and ecological sustainability of management. Subsequently, a National Forest Action Plan (NFAP) was recently adopted (1991) which calls for far-reaching policy changes and modifications at all levels of management. The overall objective of NFAP is to achieve sustainability of forest resource utilisation and to provide adequate protection of Papua New Guinea's rich biological diversity. To achieve the latter, the plan calls for the establishment of a network of World Heritage Areas, which, if realised, will raise the TPA coverage to about 20% of total national territory. Successful implementation of the NFAP will require substantial financial assistance from the international community.

Basic Forest Statistics: Papua New Guinea

Total Land Area:

46,200,000ha	(Saulei, 1990a)
45,286,000ha	(World Resources Institute, 1990. p.269)
45,171,000ha	(Collins *et al*. 1991. p.174)
46,886,000ha	(Burgess, 1988. p.6)

Total Forest Area:

36,200,000ha	(Saulei, 1990a)
38,270,000ha	(WRI, 1990. p.269) ("Forest and Woodland: 1985-87")
38,175,000ha	(WRI, 1990. p.293) ("Extent of Forest and Woodland, 1980s")
36,675,000ha	(Collins *et al*. 1991. p.174) ("Rain Forests" and "Monsoon Forests")
35,990,000ha	(Burgess, 1988. p.6)

Deforestation Rate:

0.1%	(WRI, 1990. p.293) (1980s)
0.03%	(derived from: Collins *et al*. 1991. p.174)

Production Forest Estate:

none	(Saulei, 1990a)
none	(WRI, 1990 p.293) ("Managed Closed Forests: 1980s")
6,200,100ha	(Collins *et al*. 1991.) ("Total Concessions Area") (Land in PNG is privately held and there are no state-owned production forest areas)

Totally Protected Area:[1]

924,000ha	(Saulei, 1990a)
55,000ha	(WRI, p.293) ("Protected Closed Forest: 1980s")
7323ha	(WRI, 1990. p.301) ("All Protected Areas")
29,016ha	(IUCN, 1990. p.)
986,900ha	(Collins *et al*. 1991)

Number of Units:

3	(WRI, 1990. p.301)
5	(IUCN, 1990. p.145)

---

1   Some of the discrepancy in the data here can be explained by the fact that most of the TPAs in PNG are smaller than 5000ha and are therefor not included in some references such as the *UN List* (IUCN, 1990).

Protected Areas: Papua New Guinea

1. Declared Parks

Varirata National Park	CP	1063ha
McAdam National Park	MP	2,080ha
Mt. Wilhelm National Park	WHP	4,856ha
Jimi Valley National Park	WHP	4,180ha
Mt. Gahavisuka Prov. Park	EHP	77ha
Nanuk Prov. Park	ENBP	1ha
Baiyer River Sanctuary	WHP	740ha
Moitaka Wildlife Sanctuary	NCD	43.83ha
Talele Island Nature Reserve	ENBP	1.4ha
Namanatabu Nature Reserve	CP	27.44ha
Cape Wom Memorial Park	ESP	105.42ha
Paga Hill Scenic Reserve	NCD	13.12ha

2. Approved Areas

Kokoda Trail	OP	93.7km
Afore Lakes	OP	54.42ha
Mt. Susu Nature Reserve	MP	49ha
Ramu National Park	MADP	64,100ha
Mt. Brown & Maria	CP	NK
Waigani Swamps	NCD	1380ha

3. Proposed Areas

Mt. Busave National Park	SHP	2500ha
Mt. Giluwe National Park	SHP	2023ha
Mt. Kemenagi Prov. Park	SHP	992ha
Lake Dakataua	WNBP	4920ha
Lake Hargy	WNBP	880ha
Embi Lakes	OP	8700ha
Lelet Plateau	NIP	NK
Mt. Karimui	SP	NK
Idlers Bay	NCD	25 AC=10.13ha
Sialum Terraces	MP	NK
Fly Islands	MP	NK
Mt. Michael	EHP	3000ha

4. Declared Management Areas

Iomare	CP	3827.5ha
Mojirau	ESP	5079ha
Neiru	GP	3984ha
Balek	MADP	470ha
Crown Island	MADP	58,969ha
Nerolowa	ManP	5850ha
Mt. Kaini	MorP	1502.835ha
Mt. Nuserang	MorP	22.23ha
Tirung Island	NSP	43,200ha
Siwilltame	SHP	12,540ha
Naza	WP	184,230ha

5. Proposed Management Areas

Mt. Otto, Kutuni Valley	EHP	2300ha
Ebota Crater Mt.		
Ubai Gubi Mts.	EHP	1000km^2
Bundi-Simku	MADP	28km^2
Enoweti	MADP	15,000ha
Naru Bench Mark	MADP	975ha
Nabi Wildlife	MADP	3ha
Low Island	MANP	16ha
Atzera Foothills	MORP	600ha
Liga Egg Ground	NIP	15ha
Echuda Tatch	NIP	6000m^2
Hambareta	NP	11,560ha
Kumusi Res. Areas	NP	291ha
Lake Kutubu W/A	SHP	660ha
Mt. Bosavi	SHP	370km^2
Mt. Giluwe	SHP	240ha
Nadobaing	WPT	400km^2
Vanimo	WSP	80,000ha

Bibliography

Burgess, P.F. 1988. Natural forest management for sustainable timber production: the Asia/Pacific region. Unpublished report prepared for IIED and ITTO.

Collins, N.M., Sayer, J.A., and Whitmore, T.C. 1991. *The Conservation Atlas of Tropical Forests: Asia and the Pacific*. Macmillan Press Ltd., London. 256pp.

Eaton, P. 1986. Tenure and taboo: customary rights and conservation in the South Pacific. Unpublished paper. University of Papua New Guinea.

IUCN. 1975. Proceedings from the South Pacific conference on national parks and reserves in Suva, Fiji. IUCN, Gland, Switzerland.

IUCN. 1990. *1990 United Nations List of National Parks and Protected Areas*. IUCN, Gland, Switzerland and Cambridge, UK.

Saulei, S.M. 1990a. Forest Resource Conservation in PNG, Unpublished report prepared for the IUCN Forest Conservation Programme workshop, "Realistic Strategies for Tropical Forest Conservation" in Perth, Australia.

Saulei, S.M. 1990. Constraints for economic development and conservation of Papua New Guinea tropical forests. A paper presented and the fifth International Ecology Congress Meeting, 23-29 August in Yokohama, Japan.

World Resources Institute (WRI). 1990. *World Resources: 1990-91*. Oxford University Press. Oxford. 383pp.

PAPUA NEW GUINEA
TOTAL FOREST AREA

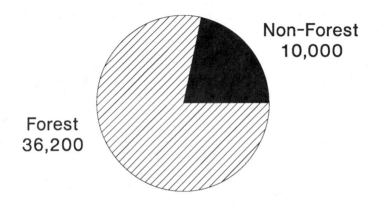

Non-Forest
10,000

Forest
36,200

Note: values given in 1000s of ha

PAPUA NEW GUINEA
LAND USE DESIGNATIONS

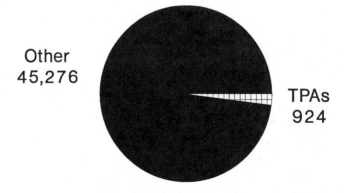

Other
45,276

TPAs
924

Note: values given in 1000s of ha

PHILIPPINES

Based on the work of Cesar Nuevo

1. Conclusions and Recommendations

Several promising initiatives have been prepared, including an integrated protected areas system for Palawan and a Master Plan for forest management. Active adoption of the Master Plan, incorporating forest environmental management, education and research, in addition to improvement of strategies in forest industry and marketing, would greatly ameliorate the status of national forest management. Given the serious state of the Philippine forests, undeniably the worst in tropical Asia, these initiatives are critical and require priority attention to implement them.

2. Forest Reserve Management and Trends

The Department of Environment and Natural Resources (DENR) has the vested authority to protect and manage forests, by regulating timber concessions and limiting kaingin (shifting cultivation) activities. DENR is constrained by its small number of staff relative to the area to be monitored and the lack of training and authority with which to carry out effective conservation.

The concession system allows for a lease of up to 25 years. The average period for which concessions are held, however, is 5-10 years. Very few timber concessions practice proper selective logging techniques or honour their regeneration requirements. It is intended that the concession system will eventually be superseded by timber production sharing agreements whereby the government is a major partner and the standing timber is sold at current market rates.

Because of the lack of sound management strategies for defining annual allowable cuts and maintaining quality harvesting operations, sustained yield management (SYM) falters. Timber exports have declined as forests are converted to agriculture or are degraded by logging.

The selection of production forests has largely been determined by the requirements of timber concessionaires (Burgess, 1988).

Fees charged do not reflect the rate of resource deletion, nor do they cover the costs associated with replanting or practicing SYM methods. Good silvicultural techniques were developed in the Philippines, such as the Philippine Selective Logging System. These techniques are, however, rarely applied. The residual stands are damaged by poor log extraction techniques. The maintenance of the forest estate is threatened by poor logging practices followed by agricultural clearance.

Despite a ban on log exports since 1989, timber smuggling is widespread. Logs are often taken to neighboring countries, such as Malaysia, for processing before being shipped further.

3. Extent, Status and Security of TPAs

The total land area of the Philippines is 29,817,000ha, of which 6,692,700ha is forested. There are 0.99 million ha of primary forests, 3.8 million ha of secondary forest and the remaining area is composed of mangroves and pine forests. Together, these forests cover 20% of national land area. This represents a large decline from the 15 million ha of primary forests which covered 50% of the national land area during the 1960s.

According to the *1990 United Nations List of National Parks and Protected Areas*, there are 28 TPAs in the Philippines covering 583,999ha. This includes only some aquatic areas and only units over 1000ha. Various national authorities are in disagreement as to the exact number of national parks [e.g. 62 (NRMC, 1983), 59 (Haribon Foundation/DENR, 1988) and 72 (Petocz, 1988)]. Regardless of the extent of parks, it is apparent that areas under forest reserves are insufficient and their protection role is inadequate. The Haribon Foundation (1986) verified that the international standards, set by IUCN, could not be met by any of the national parks. The Master Plan stated the situation to be critical. Considerable encroachment into these areas has been observed, but the extent has never been appraised. Among the most threatened protected areas, Mt. Apo National Park, is degraded by human settlement, logging and removal of vegetation. In 1985, a portion of the Mt. Apo National Park was declassified for agricultural development.

In an attempt to limit intensified kaingin practices, social forestry programmes have been introduced. Under the Integrated Social Forestry Programme (ISFP), families are granted tenure to land which they have occupied for more than 25 years, as an incentive to encourage sustainable agricultural practices.

In the Philippines, all land not alienated is declared to be forest land. The implications of such a policy are biased against establishing forest reserves, and often the forests are merely residual state land which may not even be forested. The legal basis for the establishment of protected areas is outdated, fragmented and complex. Since the turn of the century, 262 enactments regarding establishment or modification of protected areas have been filed. This has resulted in a system which is redundant, inconsistent and is lacking adequate parameters to delimit protected areas. Poor definitions and criteria for protected areas have led to a proliferation of reserves lacking proper planning for their integrity and management. An estimated two-thirds of the national parks contain human settlements and report significant alteration of the original vegetation.

Recognition of the poor conservation management of the protected areas system has led to the formulation of the Integrated Protected Areas System (IPAS). This system has a mandate to enlarge existing sites and demarcate critical sites for conservation and protection. In essence, it would reduce the amount of terrestrial reserves from 59 to 28, while increasing the area protected from 410,000 ha to 645,000 ha (Haribon Foundation/DENR, 1988). The system would be seriously altered as only 19 of the 59 National Parks would retain their status. The other 40 would be relinquished to local control by municipal or provincial governments or the Department of Tourism to be managed as recreational areas for the benefit of the local community (IUCN, 1991).

The Integrated Protected Areas System initiative would include legislative reform, institutional changes, international support for identifying and protecting areas of rich biodiversity, site-specific management and a rigorous programme of conservation education.

4. Extent, Status and Security of Production Forests

Burgess (1988) estimated that of the total forest area of 6,383,000ha; permanent protection forests amount to 1,680,000ha, and production forests total 4,403,000ha. The World Conservation Monitoring Centre (1988) reported that only 0.7% of the country's land area is designated as protected and that a mere 0.3% more is proposed for protection. No protection exists for freshwater swamps, mangroves, pine forest and the areas of lowland forests protected are "pathetically small".

The Master Plan is a 25 year strategy for restoration of degraded forests and management of intact forests. Its main objectives are:

1) local community participation in reforestation schemes;

2) private sector investments in industrial plantations;

3) cultivation of non-timber products;

4) protection and rehabilitation of critical watersheds; and

5) re-evaluation andconsolidation of the protected areas system.

The Master Plan, once given high-level profile and priority by the GOP will contribute decisively to the improved management of not only the protected areas system, but the entire permanent forest estate.

Basic Forest Statistics: Philippines

Total Land Area:
29,817,000ha   (World Resources Institute, 1990. p.269)
29,817,000ha   (Collins *et al.*, 1991. p.192)

Total Forest Area:
6,692,700ha   (Nuevo, 1990)
11,150,000ha   (WRI, 1990. p.269) ("Forest and Woodland: 1985-87")
9,510,000ha   (WRI, 1990. p.293) ("Extent of Forest and Woodland, 1980s")
6,602,000ha   (Collins *et al.*, 1991. p.192) ("Rain Forests" and "Monsoon Forests")
6,383,000ha   (Burgess, 1988. p.6)

Deforestation Rate:
1.5%   (WRI, 1990. p.293) (1980s)
1.85%   (derived from: Collins *et al.*, 1991. p.192)

Production Forest Estate:
4,403,000ha   (Burgess, 1988. p.6)

Watershed Protection Forests:
1,680,000ha   (Burgess, 1988. p.6)

Totally Protected Area:
583,999ha   (IUCN, 1990. p.147)
690,000ha   (WRI, 1990. p.293) ("Protected Closed Forest: 1980s")
520,816ha   (WRI, 1990. p.301) ("All Protected Areas")
565,600ha   (Collins *et al.*, 1991. p.199)

Number of Units:
32   (WRI, 1990. p.301)
28   (IUCN, 1990. p.147)

Bibliography

Burgess, P.F. 1988. Natural forest management for sustainable timber production: the Asia/Pacific region. Unpublished report prepared for IIED and ITTO.

Collins, N.M., Sayer, J.A., and Whitmore, T.C. 1991. *The Conservation Atlas of Tropical Forests: Asia and the Pacific*. Macmillan Press Ltd., London. 256pp.

Cox, R. 1988. The conservation status of biological resources in the Philippines. An unpublished report prepared by WCMC for IIED, Cambridge, UK.

IUCN. 1988. *The Conservation Status of Biological Resources in the Philippines*. IUCN Conservation Monitoring Centre, Cambridge, UK.

IUCN. 1990. *1990 United Nations List of National Parks and Protected Areas*. IUCN, Gland, Switzerland and Cambridge, UK.

Haribon Foundation. 1986. Assessment and study of National Parks: a proposal. Unpublished document. 16pp.

Haribon Foundation/DENR. 1988. *Development of an Integrated Protected Areas System (IPAS) for the Philippines*. World Wildlife Fund-US, Department of the Environment and Natural Resources, Department of Natural Resources, Haribon Foundation, Manila, Philippines.

NRMC. 1983. *An Analysis of Laws and Enactments Pertaining to National Parks*. Volume One. Natural Resources Management Centre, Quezon City, Philippines.

Nuevo, C. 1990. Conservation efforts and biodiversity awareness in the Philippines. Unpublished paper prepared for the IUCN Forest Conservation Programme workshop on "Realistic Strategies for Tropical Forest Conservation" in Perth, Australia.

Petocz, R. 1988. *Philippines. Strategy for Environmental Conservation*. A draft report to WWF-US and Asian Wetland Bureau. 66pp.

WCMC/IUCN Tropical Forest Programme. 1988. *Philippines: Conservation of Biological Diversity and Forest Ecosystems*. WCMC, Cambridge, UK.

World Resources Institute (WRI). 1990. *World Resources: 1990-91*. Oxford University Press. Oxford. 383pp.

PHILIPPINES
TOTAL FOREST AREA

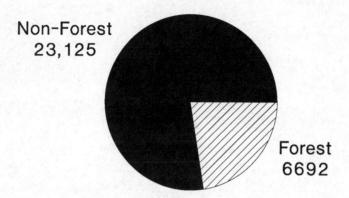

Non-Forest
23,125

Forest
6692

Note: values given in 1000s of ha

PHILIPPINES
LAND USE DESIGNATIONS

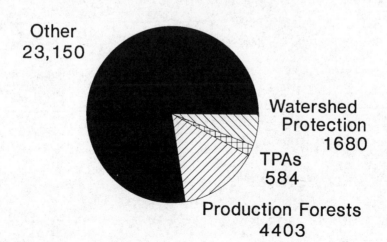

Other
23,150

Watershed
Protection
1680

TPAs
584

Production Forests
4403

Note: values given in 1000s of ha

THAILAND

Based on the work of Suvat Singhapant
Prepared by IUCN Staff

1. Conclusions and Recommendations

Thailand has embarked upon a policy whereby natural forest areas will eventually only persist inside TPAs. Some secondary forests will persist outside TPAs but these are either being cleared for agriculture or converted to industrial plantations.

The TPA system is extensive but has suffered much disturbance and does not contain significant examples of the most important forest type for biodiversity; lowland evergreen and semi-evergreen forest.

There are possibilities for bringing some secondary forest under sustained yield management and this would contribute to biodiversity conservation. This is being considered in the context of the Forestry Master Plan which is at present under preparation. It contains provision for the examination of natural forest management as one option for the forestry sector in Thailand. ITTO should encourage this initiative.

2. Extent, Status and Security of TPA System

Sixty-three National Parks cover a total of 3,386,700ha. Parts of 14 units cover coastal water areas. An additional 45 units covering 2,261,700ha are proposed. Thirty-two Wildlife Sanctuaries cover a total of 2,495,000ha, a further 6 units covering 219,600ha are proposed. National Parks and Wildlife Sanctuaries are subject to roughly equivalent conservation management regimes except that visitation is promoted in parks and only allowed on a limited scale around nature education centres in the sanctuaries. These two categories of TPAs will eventually cover 16.3% of the country. The only other protected area category of any significance for forest conservation are forest parks, but these only cover a very small area of forest.

Few of the protected areas are at present managed to international standards. Illegal logging, poaching and settlements are widespread. The latter is a particularly serious threat to the forests of the uplands in the north of the country where ethnic minority hill tribes practise low-grade agriculture in many TPAs. In the past, policy has always been to gazette as many areas as possible, even when there was little prospect of being able to apply satisfactory management regimes. The hope was that in the future increased resources and increased public support would enable improvement in management to be brought about.

This policy appears to be valid as the Royal Forest Department does now have somewhat greater resources and is gradually attempting to strengthen its control over most TPAs. Public support for conservation is evolving rapidly and several NGOs are now active. The media, and increasingly some politicians, are paying much more attention to conservation than they did in the past. However, some TPAs have suffered developments and permanent settlements and it is now necessary to rationalise their boundaries.

FAO (1981) and IUCN (1986) have evaluated TPA coverage. A major deficiency lies in the almost total absence of any examples of lowland evergreen or semi-evergreen forest in the TPAs. This formation, which is of particularly great importance for biodiversity, has virtually all been cleared for agriculture or tree crops. Many TPAs were previously National Forest Reserves and were logged. Regenerating logged-over forest occupies considerable areas in some parks and sanctuaries.

3. Extent, Status Security of Production Forests

Government policy is to achieve 41% forest cover in the Kingdom. At present it is officially estimated that 28% of the land is under forest. All logging of natural forest has been banned since early 1989, although this ban is at present being challenged in the courts by concessionaires. Most natural forest outside TPAs has suffered serious disturbance in the past. There appears to be little correspondence between location of forest and location of National Forest Reserves (IUCN 1982). The government is providing powerful incentives to the private and corporate sector to invest in plantation forestry. NFR land is rented at inferior rents (2 US cents/ha) for plantation purposes. The land rented is often covered in secondary forest which is cleared for the plantations.

The situation is thus tending towards one in which natural forest will only occur in TPAs (16.3% of territory) and plantations are replacing all other forest types and in addition are being established on deforested land and are destined to constitute the balance of the target of 41% forest cover. At present rates of plantation establishment (40,000ha per year) and assuming all deforestation could be halted, this would take 80 years. An estimated 234,500ha is deforested each year, five times greater than the plantation area established.

A recent study by the prestigious Thailand Development Research Institute (TDRI) has strongly criticised government policy which encourages replacement of natural forest by plantations and has suggested that all natural forest should be protected. A Forestry Master Plan is at present being prepared with assistance from FINNIDA. No indication of the direction that it will adopt is yet available but fears have been expressed that it may be too oriented towards industrial plantation forestry.

4. Management of Production Forests and Implications for Biodiversity Conservation

No natural forests are at present under management for timber. One consequence of this is that Thailand will in future depend on imports for all its high quality veneer and joinery timber (except teak). Much of the market for these is at present met by quasi-legal imports from Myanmar (Burma), Laos and Cambodia.

Considerable areas of land are covered with degraded forest. Much of this land has little agricultural potential and it is not required to compliment the existing TPA system. The optimum use of this land would probably be sustained yield selective logging and harvesting of non-timber products. Such a permanent production forest estate might cover between 5 and 10% of national territory. It could yield a variety of products which are essential to the Thai economy, particularly some of the high value timbers which are used for quality furnishings. Teak from natural forest and rosewoods (*Dalbergia* and *Pterocarpus*) could be harvested from these forests.

It is in any case quite unrealistic to think that the government target of 41% forest cover can be met through plantations. There is not sufficient land with the potential to justify the heavy investments required to establish and maintain industrial plantations. If there were, there would not be markets for the vast amount of low-grade lumber that would thus be produced.

It is therefore recommended that ITTO promote the concept of greater emphasis being given to the management of natural forests in marginal areas in Thailand.

Basic Forest Statistics: Thailand

Total Land Area:

51,089,000ha	(World Resources Institute, 1990. p.269)
51,117,700ha	(Collins *et al*. 1991. p.222)

Total Forest Area:

14,313,000ha	(derived from unpublished data from the Royal Thai Forestry Department)
14,662,000ha	(WRI, 1990. p.269) ("Forest and Woodland: 1985-87")
15,675,000ha	(WRI, 1990. p.293) ("Extent of Forest and Woodland, 1980s")
10,690,000ha	(Collins *et al*. 1991. p.226) ("Rain Forests" and "Monsoon Forests")
14,295,800ha	(Burgess, 1988. p.6)

Deforestation Rate:

1.6%	(derived from unpublished data from the Royal Thai Forestry Department)
2.5%	(WRI, 1990. p.293) (1980s)
2.8%	(derived from: Collins *et al*. 1991. p.222)

Production Forest Estate:

8,856,000ha	(WRI, 1990 p.293) ("Managed Closed Forests: 1980s")

Totally Protected Area:

5,881,700ha	(derived from unpublished data from the Royal Thai Forestry Department)
2,220,000ha	(WRI, 1990. p.293) ("Protected Closed Forest: 1980s")
4,676,757ha	(WRI, 1990. p.301) ("All Protected Areas")
5,105,746ha	(IUCN, 1990. p.170)
5,137,800ha	(Collins *et al*. 1991. p.229)

Number of Units:

95	(derived from unpublished data from the Royal Thai Forestry Department)
75	(WRI, 1990. p.301)
83	(IUCN, 1990. p.170)
97	(Collins *et al*. 1991. p.229)

Bibliography

Burgess, P.F. 1988. Natural forest management for sustainable timber production: the Asia/Pacific region. Unpublished report prepared for IIED and ITTO.

Collins, N.M., Sayer, J.A., and Whitmore, T.C. 1991. *The Conservation Atlas of Tropical Forests: Asia and the Pacific*. Macmillan Press Ltd., London. 256pp.

FAO. 1981. *National Parks and Wildlife Management, Thailand: A Review of the Nature Conservation Programmes and Policies of the Royal Thai Forest Department*. FAO/UNDP, Bangkok.

IUCN. 1986. *Review of the Protected Areas System in the Indo-Malayan Realm*. IUCN, Gland, Switzerland and Cambridge, UK. 284pp.

IUCN. 1990. *1990 United Nations List of National Parks and Protected Areas*. IUCN, Gland, Switzerland and Cambridge, UK.

World Resources Institute (WRI). 1990. *World Resources: 1990-91*. Oxford University Press. Oxford. 383pp.

THAILAND
TOTAL FOREST AREA

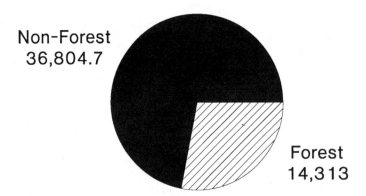

Non-Forest
36,804.7

Forest
14,313

Note: values given in 1000s of ha

THAILAND
LAND USE DESIGNATIONS

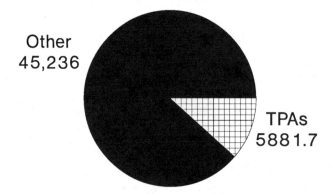

Other
45,236

TPAs
5881.7

Note: values given in 1000s of ha

ASIA OVERVIEW*

By
Sin Tuan Mok

1. Conclusions and Recommendations

The Asia-Pacific region is physically, culturally and biologically very diverse. The countries in the region are progressing politically and socio-economically at different rates as each country has its own basic set of conditions, directions and expectations, problems and constraints, and policies and strategies. Consequently, each country must ultimately find its own set of solutions based on its own specific objectives, capabilities and resources.

Despite its long history of land-based cultural and socio-economic development, the Asia-Pacific region is still relatively well-endowed with forest resources, due largely to the peoples' traditional belief in living in harmony with nature and praticising sustainable agriculture. However, with growing population and increasing demands on the forests, some countries already face a shortage of forest resources but many still have an abundance of them. In general, the countries in West, South and East Asia have a deficit while those in South East Asia and Oceania are better endowed.

The peoples' awareness of the forest's multiple functions and vital contributions to their welfare and quality of life prompted the countries in the Asia-Pacific region to establish extensive reserves and areas for environmental protection and biological diversity conservation, often for non-wood forest products to satisfy local needs. Most countries have a core network of protected areas, although representation of ecosystems may be inadequate in many countries due mainly to insufficient information on the distribution of biological diversity. While these countries acknowledge the need to enhance the existing protected areas system, many are faced with the dilemma of serious land use conflicts, which is exacerbated by inadequate managerial and technological capabilities and insufficient human and financial resources.

The countries of the Asia-Pacific region depend on the forests to varying degrees to meet the peoples' basic needs and to generate socio-economic benefits. Consequently, permanent forest estates and permanent production forests have been established to protect environmental and natural resources and to produce timber and non-wood forest products. Most of the production forests are located in natural forest areas although many countries have embarked on ambitious programmes to reforest degraded and deforested sites, which will effectively enlarge the forest resource base for timber production and hopefully reduce the pressure on the remaining natural forests. In view of the limited scope and poor prospects for expanding the protected area systems in most of the countries in the Asia-Pacific region, it is imperative that the permanent production forests are managed sustainably to ensure that biological diversity is conserved effectively. Unless sustainable management is applied to the remaining natural forest resources and new forests are created, the Asia-Pacific region will not only suffer further losses of valuable biological diversity, but more importantly the region will become a net importer of forest products.

---

*   *Some of the forest statistics used in this regional overview differ from those used in the country reports. Where differences exist, these are indicated in footnotes.*

The management of forest land has been praticised in the Asia-Pacific region for a long time to varying degrees, depending on the type and location of the forest land and the availability of human, logistical and financial resources. Remote and protected areas tend to be managed minimally due to limited human and financial resources generally available in most of the countries in the Asia-Pacific region, and the common perception that benefits from management are unlikely to be commensurate with the costs. On the other hand, some form of management is always praticised in the permanent production forests. Very often, however, management is impractical, inefficient, ineffective or even counter-productive when conventional and largely Eurocentric ideals, concepts, principles and practices are applied inflexibly and inappropriately. The practice of sustainable forest management in many countries is constrained by a lack of political will and direction, often due to a lack of awareness of the important functions of the forests amongst those responsible for national policy and strategy decisions as well as those who advise the decision makers. In general, most countries are also constrained by insufficient information on the forest resources and biological diversity; inability to quantify the characteristics and value of the biological resources; inability to detect and monitor changes in the resources in a timely manner; inappropriate or inefficient technologies for forest management, harvesting and reforestation; ineffective managerial and operational skills; and insufficient human and financial resources.

The resolution of the inherent weaknesses and gaps in the Asia-Pacific region by itself will not necessarily ensure sustainable forest management and effective biological diversity conservation. In order to attain the dual objectives of sustained yield of forest products and effective conservation of biological diversity, it is essential for a country to adopt a comprehensive and realistic forest management strategy with pragmatic programmes which accord high priority to the following:

1)   Sustainable management of the natural forests.

2)   Afforestation or restoration of degraded and deforested areas, especially in critical watersheds.

3)   Research and development targeted at more realistic policy and strategy options for forest management and development, to include among the following:

   a) the development of appropriate forest management, harvesting and reforestation technologies;

   b) the upgrading of managerial and operational skills;

   c) the generation of relevant and reliable information necessary for conservation and sustainable forest management and reforestation; and

   d) the development of the most efficient and cost-effective tools and methodologies for the acquisition, management and dissemination of such information.

4)   Public education to encourage the appreciation of natural forests and to motivate forestry staff to praticise sustainable forest management and effective biological conservation.

The conservation and sustainable management of the natural forests of the Asia-Pacific region requires the evolution of a forestry culture which is compatible with the local ecological, environmental, political, social, cultural, and economic conditions. There is also an urgent need to upgrade the essential technical, managerial and operational skills to formulate and implement "conservation forest management strategies" and pragmatic programmes, especially to intensify integrated forest resources inventory and management and integrated studies in forest management and operations designed to develop the appropriate technologies and generate relevant information. As most of the countries in the Asia-Pacific region are unlikely to be able to mobilise the necessary resources to do so nationally, it is recommended that the International

Tropical Timber Organization and development assistance organisations collaborate to establish the following:

1) An Asia-Pacific Forestry Centre (or a programme at an existing institution such as the Asian and Pacific Development Centre) to conduct policy and economic studies to evolve a regional "Tropical Forestry Culture" – an overall approach to tropical forest management, adapted to regional and local conditions. This centre would potentially evaluate realistic strategic options for sustainable forest management and effective biological diversity conservation.

2) Centres of Excellence for Sustainable Forest Management and Development to develop, adapt, package and transfer the appropriate technologies, methodologies, techniques, skills and information effectively to the countries of the Asia-Pacific region.

3) Regional forestry development programmes for Continental Asia, South East Asia, and the South Pacific, which has been proposed by the Regional Office for Asia and the Pacific of FAO, to upgrade the necessary managerial, technical, technological and operational skills to ensure sustainable forest management and development which will contribute to effective conservation of biological diversity.

2. Introduction

The Asia-Pacific region, comprising some 40 countries, is home to more than half of humankind. It is a region of geographic, climatic, demographic, socio-economic and technological extremes. The region is culturally, politically, socially, economically, environmentally and biologically highly diverse. Nevertheless, the region's indigenous religions and philosophies generally embody the concept of living in harmony with nature. Its traditions, customs and cultural activities are oriented towards the achievement of sustainable development which ensure the people's survival and quality of life.

The Asia-Pacific region has been well endowed with natural forests which satisfy the basic needs of the people and contribute to the socio-economic progress of many countries. Closed forests cover about 495.7 million ha of which only 48.7 million ha are managed and 19.5 million ha protected (WRI, 1990). The most extensive forest resource of the region is the tropical forests of South East Asia which cover some one million square kilometers (Collins *et al.*, 1991). The region has a great diversity of plant and animal species of economic importance and is one of the cradles of agriculture where many of the world's major crop plants originated or have their centre of diversity. Many of the plants and animals from the forests are important sources of fruits, drugs, oils, beverages, gums, vegetables, spices, fibres, medicinal and ornamental plants, and rattan. Although timber and rattan are presently the major commercial forest products, it is a recent phenomenon in the region as the non-wood or minor forest products have been more important both to the local rural economies as well as in the export trade up to about the middle of this century. Consequently, the countries of the Asia-Pacific region recognise the vital role of the forests and appreciate the need for sustainable development and effective conservation of biological diversity.

Most of the countries in the Asia-Pacific region have technically sound forestry policies, legislation and institutions. Nevertheless, the forestry strategies and programmes of many of these countries are often unrealistic or impractical while inter-sectoral coordination essential for sustainable development is often lacking or ineffective. Consequently, the increased efforts in socio-economic development in most of these countries have invariably contributed to an acceleration of deforestation and forest degradation. The situation is exacerbated by the rapid growth in population which demands more food and fuelwood as well as employment which is often created by the conversion of forests to commercial agricultural crops and the development of wood-based industries. The escalating rates of deforestation and forest degradation have inevitably led to loss of biological diversity and detrimental environmental impacts, both locally and, possibly, globally.

The countries of the Asia-Pacific region have acknowledged the seriousness of the problems caused by deforestation and forest degradation and are endeavouring to resolve them. Many countries have reserved extensive areas of representative forest types as totally protected areas (TPAs) for the conservation of biological diversity and for the protection of critical watersheds and other sensitive lands. Protected areas have also been established as permanent production forests which are managed in accordance with the concept of sustained yield of timber and non-wood forest products. In spite of these measures, however, the forest resource base of most of the countries continue to be eroded due to their inability to praticise sustainable management effectively. Many of the countries are constrained by bureaucratic and conservative institutions with incompatible "Eurocentric" concepts and ideals; inadequate managerial expertise and skills; inappropriate technologies and operation systems; irrelevant and unreliable information; and insufficient human and financial resources. Those countries exporting forest products are often confronted with unfavourable terms of trade; unreasonable tariff barriers; inspired consumer resistance and discrimination; and industrial subterfuge.

3. Totally Protected Areas

The forestry institutions in many countries of the Asia-Pacific region are amongst the oldest in the world. Consequently, basic forestry policies and legislation of these countries normally provide for the protection and management of the total forest estate with broad categories of forest reserves and management objectives which embrace all the IUCN "Categories and Management Objectives of Protected Areas". However, as the countries in the Asia-Pacific region have been influenced by different foreign cultures in their recent history, the terminology and interpretation of the various categories of protected areas and their management objectives will no doubt differ in detail from country to country and may not coincide with those of IUCN.

Most of the countries in the Asia-Pacific region have three major categories of forests similar to those of Malaysia which are as follows:

1) **Protection Forests**: to ensure sound climatic and physical conditions of the country, the safeguarding of water supplies, soil fertility and environmental quality, and the minimisation of damage by floods and erosion to rivers and agricultural lands.

2) **Productive Forests**: to ensure the supply in perpetuity at reasonable rates of all forms of forest produce which can be economically produced within the country and are required for agricultural, domestic and industrial purposes.

3) **Conservation or Amenity Forests**: to ensure the conservation of adequate forest areas for recreation, education, research and the protection of the country's unique flora and fauna.

In recent years, changing perceptions and expectations of forests have prompted many countries to promulgate separate policies and legislation for environmental protection and biological diversity conservation. As a consequence, national parks and wildlife reserves or sanctuaries have been established with management objectives which correspond to those of IUCN Categories II and IV. In some of the countries, separate institutions have also been established, some times under different authorities, resulting in unnecessary areas of conflict, particularly in the allocation of forest lands and their administration, which are not conducive or are even counter-productive to effective protection of ecosystems and the conservation of biological diversity.

A "totally protected area" (TPA) is generally considered to include IUCN Categories I to V. These categories and corresponding management objectives are defined as follows:

I   **Scientific Reserve/Strict Nature Reserve**: to protect nature and maintain natural processes in an undisturbed state in order to have ecologically representative examples of the natural environment available for scientific study, environmental monitoring, education, and for the maintenance of genetic resources in a dynamic and evolutionary state.

II   **National Park**: to protect natural and scenic areas of national or international significance for scientific, educational and recreational use.

III **Natural Monument/Natural Landmark**: to protect and preserve nationally significant natural features because of their special interest or unique characteristics.

IV   **Managed Nature Reserve/Wildlife Sanctuary**: to assure the natural conditions necessary to protect nationally significant species, groups of species, biotic communities, or physical features of the environment where these require specific human manipulation for their perpetuation.

V   **Protected Landscape or Seascape**: to maintain nationally significant natural landscapes which are characteristic of the harmonious interaction of man and land while providing opportunities for public enjoyment through recreation and tourism within the normal lifestyle and economic activity of these areas.

The extent of protected areas in the countries of the Asia-Pacific region has been compiled by the IUCN (WCMC, 1990). According to the IUCN report, the Asia-Pacific region has 2303 protected areas covering 139,239,136ha of which 810 areas covering 66,171,615ha are totally protected and 1493 areas covering 73,067,521ha are partially protected. The extent of protected areas in the various sub-regions is as follows:

West Asia	13,793,838ha
South Asia	20,041,867ha
East Asia	30,929,406ha
South East Asia	25,842,087ha
Oceania	48,631,938ha

Most of the countries in the Asia-Pacific region have less than 10 percent of the land protected. Countries with more than 10 per cent coverage are Kiribati (38.9%), Hong Kong (35.6%), Brunei (21.2%), Bhutan (19.8%), Sri Lanka (11.9%), Israel (10.9%) and New Zealand (10.7%). However, countries with the most extensive coverage of protected areas appear to be Australia (45,654,429ha), China (21,947,104ha), Indonesia (17,799,787ha), India (13,481,148ha) and Iran (7,528,976ha). It is interesting to note that all the five sub-regions are represented in terms of countries with more than 10 per cent coverage as well as countries with the most extensive coverage.

The extent of protected areas in the Asia-Pacific region by IUCN categories is as follows:

Category I	214 sites	12,520,439ha
Category II	582 sites	53,601,785ha
Category III	14 sites	49,391ha
Category IV	1322 sites	62,434,917ha
Category V	171 sites	10,632,604ha

The IUCN report indicates that the Asia-Pacific region has 2,817,028 sq km of tropical moist/rain forests which are known to be amongst the richest ecosystems in the world in terms of biological diversity. Of the total, South East Asia accounts for 2,163,582 sq km (76.8%) while Oceania accounts for 377,266 sq km (13.4%). The major locations of tropical moist forests in the Asia-Pacific region are Indonesia (1,179,140 sq km), Papua New Guinea (366,750 sq km), Myanmar (311,850 sq km), India (288,330 sq km), Malaysia (200,450 sq km), Laos PDR (124,600 sq km), Cambodia (133,250 sq km) and Thailand (106,900 sq km).

The total area of existing protected tropical moist forests in the Asia-Pacific region is 281,370 sq km while a total area of 233,542 sq km has been proposed for protection which will eventually increase the protected areas with tropical moist forests to 514,912 sq km. Countries with the most extensive coverage of protected areas with tropical moist forests at present are Indonesia (137,875 sq km), Thailand (44,790 sq km), India (22,658 sq km), Cambodia (20,351 sq km), and Malaysia (13,263 sq km) while those with extensive proposals are Indonesia (128,108 sq km), Laos PDR (47,211 sq km), India (18,892 sq km), Malaysia (14,388 sq km) and Thailand (11,855 sq km). If all proposals are approved, the countries with the most extensive protected areas with tropical moist forests will be Indonesia (265,983 sq km), Thailand (56,645 sq km), Laos PDR (47,211 sq km), India (41,550 sq km), Malaysia (27,651 sq km) and Cambodia (25,026 sq km). In spite of this, only Brunei (20.3%), Indonesia (14.6%), Cambodia (14.1%) and Thailand (11.1%) will have more than 10 percent of the land as protected areas with tropical moist forests. On the other hand, only a few countries will have less than 10 per cent of the remaining tropical moist forests as protected areas.

The extent of existing and proposed protected areas, their apparently wide geographical distribution, and their fairly long history of establishment suggest that most, if not all, of the major ecosystems would be represented to varying degrees. However, the coverage within the respective countries of the region will tend to favour the predominant and more accessible ecosystems due to constraints of human and financial resources, ecological information, and managerial logistics and skills. In view of the continuing depletion of the forest resources of the Asia-Pacific region and the already restricted extent of natural forests in most of the countries, the prospects for achieving the desired level of protected areas for all the ecosystems within every country of the region will be limited.

The status of protected areas in the Asia-Pacific region varies from country to country as the process for their establishment may be constitutional, legal, administrative or by consensus. Be that as it may, there can be no absolute guarantee for the status of a protected area as the same process provides for the reservation and revocation of such areas. In most developing countries of the region where the population is growing rapidly and poverty is widespread, the increasing demand for food and fuelwood by the rural and local population will invariably continue to pose a threat to the forests. Where forests are still abundant but management and law enforcement are weak, encroachment for agricultural purposes and illegal logging will continue to be rampant, irrespective of the status of the forests. Forest degradation and deforestation will be inevitable unless alternative sources of energy and gainful employment are available outside the forestry sector. In the final analysis, the status of the forests, protected or otherwise, will only be assured if socially, economically, environmentally and biologically sustainable forest management and development can be demonstrated conclusively.

Despite being a leading producer and consumer of fuelwood and industrial wood, the Asia-Pacific region managed to retain a substantial area of forest land to protect the fragile environment, satisfy basic needs and conserve biological diversity. South East Asia has been the predominant producer and exporter of tropical hardwood products for more than two decades. China, India, Japan, Australia and Papua New Guinea are still relatively well-endowed with forest resources, a substantial portion of which are existing or proposed protected areas which are managed for the protection of non-wood natural resources and conservation of biological diversity.

Protected Areas in Indonesia

Indonesia, with a total land area of 192 million ha and 144 million ha[1] of forest land spread over more than 13,000 islands, has the richest biological diversity in the Asia-Pacific region. The country, which encompasses three major vegetation regions from Asia to Oceania, is estimated to have 25,000-30,000 species of flowering/seed bearing plants (4000 trees), 750 species of mammals (100 endemic), 1250 species of birds, 12,000 species of arthropods/insects, and a host of fishes, reptiles/amphibians, molluscs, and other invertebrates. Consequently, Indonesia has an extensive network of protected areas which cover 49 million ha and represent 25.5 per cent of the land area or 34.0 per cent of the forest land. The proposed protected areas, comprising 30.3 million ha of watershed protection forests and 18.7 million ha of national parks and other reserves, are distributed throughout the Indonesia archipelago and represent virtually all the 70 ecosystems within the country. The extent of national parks, nature reserves, wildlife sanctuaries and other conservation areas exceed the national target of a minimum of 10 per cent of the total area of forests.

Although the protected areas in Indonesia are technically reasonably secure, there are instances of encroachment and illegal logging which are inevitable, given the sheer size of the country and the limited human and financial resources, logistics and information available for the management and protection of the forest resources. Provisions have been made for the establishment of natural buffer zones and corridors, sanctuary areas, wildlife refuges and seed stands, but most of them have not been implemented effectively. Consequently, an Act of Parliament for the Conservation of Living Resources and their Ecosystems has been ratified recently which stipulates that buffer zones of 500m to 1000m width must be maintained for all concessions adjoining the protected areas. Environmental Impact Assessments are mandatory for existing and proposed timber concessions as well as tree plantations being established as industrial timber estates. These measures and the more stringent requirements introduced for the preparation of forest management plans are expected to help safeguard the protected areas and promote biological diversity conservation more effectively.

Protected Areas in Malaysia

Malaysia has a land area of 32.9 million ha comprising 13.2 million ha in Peninsular Malaysia and 19.7 million ha in Sabah and Sarawak located on the island of Borneo. The biologically rich and varied tropical rain forests of Malaysia have some 14,500 species of flowering plants (890 reaching 45 cm dbh), over 800 species of non-flowering plants, around 1000 species of vertebrates, and 20-80 thousand invertebrate species. A high degree of endemism ranging from 30 per cent of all tree species to 80-90 per cent of some families has been observed. The total forest area as of 1989 was 18.4 million ha or 56.0 per cent of the land area. Dipterocarp forests covered 15.9 million ha or 86.4 per cent of the forest land. Since the 1930s, an extensive network of protected areas comprising national parks, wildlife reserves/sanctuaries, nature parks, bird sanctuaries and marine parks has been established covering 1.39 million ha of which 1.06 million ha are outside the Permanent Forest Estate. An additional area of 1.42 million ha

---

1   This figure represents legally recognised "forest land". Actual forest coverage is significantly less. See the Indonesia country summary for various estimates of forest extent.

has been proposed as national parks and wildlife sanctuaries with priority for the protection of unique habitats such as wetlands, open lakes, quartz ridges and limestone formations. A network of Virgin Jungle Reserves, established from the 1950s, served as permanent nature reserves and natural arboreta, controls for comparing harvested and silviculturally treated forests, undisturbed natural forests for ecological and botanical studies, and conservation areas for biological diversity. A total of 71 sites covering 21,272ha were located in Peninsular Malaysia with 48 sites covering 88,299ha in Sabah. Many of the Jungle Reserves are seriously depleted and retain little conservation value.

As Malaysia is a federation, the reservation and revocation of the Permanent Forest Estate and conservation areas are affected by state legislation. However, technical advice and guidance are provided by the relevant federal authorities while the harmonisation of state legislation and policies are achieved through the National Land Council and the National Forestry Council. The National Forestry Policy and the National Forestry Act provide the basis for systematic forestry development as the latter stipulates the preparation of forest management plans and the classification of forests for various functions including environmental protection and biological diversity conservation. The provisions in the forestry legislation for environmental protection and biological diversity conservation are augmented by the Protection of Wildlife Act, the Environmental Quality Act which prescribes Environmental Impact Assessments for activities involving forest land, and the National Park Act. Being a relatively small and compact country, management and enforcement in Malaysia have been reasonably effective although there are isolated cases of encroachment and illegal logging, especially in the more remote and inaccessible areas. The main constraints and problems to more effective conservation, management and development of the forest resources are the lack of security of tenure for the Permanent Forest Estate, the absence of a land-use master plan, inadequate information on the economic values of the natural biological resources, and the costs of conserving biological diversity.

Protected Areas in the Philippines

The Philippines has a total land area of 30 million ha spread over 7111 islands. Forests covered 17 million ha in the 1930s but the area has been depleted rapidly by encroachment to only 6.69 million ha[1] or 22.3 per cent of the land area by 1990 of which less than a million ha are virgin or old growth Dipterocarp forests. Protected areas cover less than two per cent of the land area comprising 215,753ha of national parks, 21,050ha of natural monument/natural landmark, 324,643ha of nature reserve/wildlife sanctuary, and 22,553ha of protected landscape/seascape (WCMC 1990). The Philippine archipelago has a unique flora and fauna with high levels of endemism in both plants and animals (FAO 1989). More than 950 terrestrial vertebrate species and some 8000 species of flowering plants have been recorded. The forests support at least 12,000 species of plants of which about 3500 are endemic; endemism being estimated at 27 per cent, with 33 endemic genera. The fauna is characterised by a high degree of endemism with about 96 species of non-volent land mammals of which at least 70 are found nowhere else in the world. The Philippine avifauna and herpetofauna, with some 860 species, also show a remarkable degree of endemism.

Conservation in the Philippines is generally poor with at least two-thirds of national parks containing human settlements and much of their total land surface supporting disturbed

---

1   6 million according to Nuevo (1990)

vegetation or agriculture (FAO 1989). The legal basis for protected areas is complex with numerous enactments, some of which overlap or are inconsistent, providing insufficient information for the protected area to be accurately demarcated. The confused legislation and administrative background and the lack of clear definition and criteria for selecting areas led to a proliferation of national parks but provided little effective protection or management for the protected areas. Most park boundaries are not demarcated and law enforcement is lacking or inadequate as current staffing and financial provisions are insufficient to deploy an effective corps of forest guards and park rangers. Consequently, proposals have been formulated to reduce the number of reserves but expand the park system by enlarging the existing sites and demarcating critical sites for conservation and protection to create an Integrated Protected Area System covering some 645,000ha.

Protected Areas in Thailand

Thailand has a total land area of 51.31 million ha of which 14.38 million ha[1] or 28.0 per cent are forest land comprising mainly Tropical Evergreen Forest (6.23 million ha), Mixed Deciduous Forest (3.12 million ha) and Dry Dipterocarp Forest (4.50 million ha) (Mok 1990). The country, which divides naturally into six geographical regions, has 10,000 to 15,000 species of plants, including 500 tree species and about 1000 species of orchids; over 900 species of birds, including 578 resident forest species of which 87 are thought to be endangered; 265 species of mammals; and 100 species of amphibians (FAO 1989). Currently, 5,881,700ha are legally protected. This includes 63 national parks covering 3,386,700ha with proposals for an additional 45 sites covering 2,261,700ha and 32 wildlife sanctuaries covering 2,495,000ha with proposals for a further six sites covering 219,600ha (derived from unpublished data provided by the Royal Thai Forestry Department).

The legal basis for conservation of biological diversity in Thailand is provided by the Wildlife Protection and Reservation Act and the National Park Act (FAO 1989). The protected area system is extensive but is far from adequate as a disproportionately large area of upland forests is protected while lowland evergreen forests are scarcely represented. With only 4.7 per cent of the lowland rain forests remaining at the end of 1985, it is likely to be too late to incorporate any significant example of this habitat into the protected area network. Furthermore, some of the protected areas are not effectively managed due to lack of public awareness and support and inadequate human and financial resources. Consequently, some of the forests in the parks and sanctuaries have been subjected to swidden agriculture while others have been logged before the areas were gazetted as reserves. Encroachment into protected areas for the cultivation of permanent crops has also been rampant and extensive, particularly in recent years. Following the ban on logging in 1989, plans have been formulated to gazette a further 15 national parks and many more sites have been proposed as national parks and wildlife sanctuaries. Public and political support for conservation is growing rapidly in response to the many recent natural disasters. The Asian Development Bank is supporting the preparation of a National Forestry Master Plan which will give special attention to ecosystem conservation issues while USAID is providing assistance to implement a programme to conserve biological diversity.

---

1   14,313,999ha according to unpublished data provided by the Royal Thai Forestry Department.

Protected Areas in Papua New Guinea

Papua New Guinea has a total land area of 46.9 million ha of which about 97 per cent is held under customary ownership (Collins *et al.*, 1991). Although approximately 20 per cent of the land area is currently used for agriculture of which 10 per cent or 4.6 million ha is under intensive cultivation, only a total of 496,000ha or one per cent is considered suitable for agriculture with no limitations. The forested area was estimated at 36.2 million ha[1] or 77 per cent of the land area in 1985. The predominant ecosystems are the Lowland Rain Forest (11.1 million ha) and the Lower Montane Forest (7.6 million ha), both of which have been subjected to shifting cultivation for thousands of years. There are also substantial areas of savanna and swamps (4.6 million ha).

The country has an extraordinary diversity of ecosystems, from mountain glaciers to humid rain forests, and some of the most remarkable wildlife on earth. New Guinea, the largest and highest tropical island, contains its own centres of endemism while the off-shore islands have their own complement of unique species. The total number of vascular plants in Papua New Guinea is believed to be in the order of 11,000 species plus about 2000 species of ferns, with estimates of endemism ranging from 55 to 90 per cent. The lowland forests are the richest with over 1200 species of trees but Dipterocarps are only locally common, and absent from large areas. Biological diversity declines with rising altitude but endemism increases. The majority of the fauna is Indomalayan in origin, with a strong Australasian influence in the mammals which has been estimated at almost 200 species. The avifauna is one of the richest and most varied in the world as New Guinea is a major centre of diversity for several birds. Approximately 740 species have been recognised, 10 per cent of them endemic, of which 445 dwell in rain forests. There are 90 species of snakes, 170 species of lizards, nearly 200 species of frogs and 455 species of butterflies over 80 per cent of which are endemic. Protected areas currently cover only 926,800ha[2] or two per cent of the land area comprising three national parks, two provincial parks, two memorial/historic parks, and one sanctuary established under the National Parks Act; and 16 wildlife management areas, three sanctuaries and one protected area designated under the Fauna (Protection and Control) Act.

The extent of protected areas in Papua New Guinea is grossly inadequate for a land of such outstanding biological diversity but the prospects for increment are poor as the complex land tenure system mitigates against major extensions. While traditional beliefs and customs have helped to protect the environment in the past, the integrity of the environment is under increasing threat from pressures associated with population growth, increased mobility and growth of the cash economy. In addition to the Fauna (Protection and Control) Act and the National Parks Act, Papua New Guinea has a Conservation Areas Act with similar objectives which provides for the establishment of a National Conservation Council to advise on the identification and management of protected areas, and the formation of management committees for each area to be responsible for *inter alia* the preparation of management plans. Although Papua New Guinea is not yet party to any of the international conventions or programmes that directly promote the conservation of natural areas, it participates in the South Pacific Regional Environmental Programme which has launched an action strategy for protected areas. Principal goals of the

---

1   35,200,000ha according to Saulei (1990)

2   924,000ha according to Saulei (1990)

strategy cover conservation education, conservation policies, establishment of protected areas, effective protected areas management, and regional and international cooperation.

Protected Areas in Nepal

Nepal is a small mountainous country with a total land area of 14.7 million ha of which 5.5 million ha[1] or 37.4 per cent is forested land. Protected areas comprising national parks and wildlife reserves cover 1.1 million ha or 7.4 per cent of the land area of which 5.7 million ha or more than half of the total forested land are located in the High Himal physiographic zone. The remaining accessible forests, comprising mostly mature stands of conifers and hardwood species, are highly degraded as a consequence of the rapidly growing population and its demand for fuelwood, timber, fodder and land for agriculture. The damage and destruction of the forests have resulted in degraded environmental quality and loss of biological diversity.

The Master Plan for the Forestry Sector of Nepal, completed in 1989 with conservation as a central theme, forms a pillar of the recently developed National Conservation Strategy. It fits naturally into the framework for the conservation of biological diversity and includes programmes designed to perpetuate the forest resources which will ensure the conservation of the diversity of flora, fauna and ecosystems in Nepal. The primary development programme for "Conservation of Ecosystems and Genetic Resources" will contribute directly to biological diversity conservation while the programme for "Medicinal and Aromatic Plants and Other Minor Forest Products" will contribute indirectly. Projects are being prepared to strengthen capability for the effective management of protected areas, the preparation and implementation of management plans, and the creation of a computerised database of plant genetic resources. The King Mahendra Trust for Nature Conservation is actively involved in conservation projects which promote the concepts of multiple land use and buffer zones. A National Commission for the Conservation of Natural Resources exists but is ineffective as its mandate is restricted to activities undertaken in watersheds.

Protected Areas in India

India has a total land area of 297.3 million ha of which 121.5 million ha[2] was classified as forest and woodland (WRI 1990). In spite of its huge population and limited land and forest resources, the country has a significant protected areas system with a total of 359 sites covering 13,481,148ha or 4.3 per cent of the land area. Of the total, two sites covering 196,043ha are classified as scientific reserve/strict nature reserve, 57 sites covering 3,329,300ha as national park, 299 sites covering 9,937,205ha as managed nature reserve/wildlife sanctuary, and 1 site covering 18,600ha as protected landscape/seascape.

The protected areas in India suffer from considerable encroachment and degradation. India has been fairly successful in conserving tigers, while at the same time addressing the problem of attacks on people and livestock.

---

1   6,224,000ha according to Ministry of Forests and Soil Conservation (1988)

2   64,013,400ha according to Collins (1991)

4. Production Forests

The concept of a permanent forest estate or permanent production forest is a forestry ideal which is widely and frequently advocated but is seldom achievable or sustainable in developing countries with large and rapidly growing populations. The most pressing immediate goals of these countries are invariably the satisfaction of the peoples' basic needs from the forest and the generation of socio-economic benefits from land-based development to meet rising expectations. This is evident in the Asia-Pacific region from the high level of deforestation which was estimated at 4.7 million ha per year during the 1980s and the total roundwood production of more than 1001 million m^3 during 1985-87 (WRI 1990).

The concept of permanency is unrealistic and illogical as the implied static state is unacceptable in a dynamic society, in which needs and priorities are changing rapidly over time. A forest estate can only be justified if the existing forest resource can satisfy the peoples' current needs and the demand to convert forests with no perceived or immediate values to other forms of land use with higher economic values will be inevitable. Forests can only be retained permanently if they are shown to be valuable or indispensable, managed with specific objectives which reflect both current and inherent values, and utilised in accordance with the concept and principles of sustainable development.

The establishment of a permanent forest estate or permanent production forest seems to be the idealistic goal of many of the countries in the Asia-Pacific region, but only a few of them have progressed beyond the conceptual stage. This could be due to a lack of consensus on the concept, which is often not well defined or clearly understood, or conflict in land use, which is very common as most of the countries in the region seem to lack rational and sound land-use policies and strategies. Despite the importance of forests as an economic resource in most of the countries in the Asia-Pacific region and forestry as a vital economic sector in some of them, most of the countries in the region still classify their forests into vegetation and ecological types with little or no indications of their resource values. While some countries have further classified their forests into protective, production and amenity groups, there seems to be no clear definition for production forests apart from vague and qualitative references to physical constraints due to slope, elevation and accessibility. Consequently, the extent, status and security of the permanent forest estate or permanent production forest in the Asia-Pacific region tend to be variable and uncertain.

Production Forests in Indonesia

Indonesia has the most extensive area of closed forest in the Asia-Pacific region and its area of managed closed forest in the 1980s exceeded the total area for all the rest of the countries in the region. Based on "Forest Land Use by Consensus", Indonesia has designated 64.4 million ha[1] as permanent production forest, which is 33.5 per cent of the total land area or 44.7 per cent of the forest area. Of the total, 33.9 million ha or 52.6 per cent are unlimited production forest where clear cutting is allowable, if necessary, and 30.5 million ha or 47.4 per cent are limited production forest where only selective cutting is allowed. In addition to the permanent production forest, Indonesia has designated 30.5 million ha as convertible production forest which will be logged and the land can be used for the development of agricultural and industrial plantation crops.

---

1   Much of this area is not forested. See the "Basic Forest Statistics" section of the Indonesia country report for various estimates of actual forest cover.

Besides the natural forest resources, Indonesia is committed to an ambitious and aggressive policy of reforestation, rehabilitation and the development of highly productive industrial timber estates by the private sector. The strategy includes the establishment of 1.5 million ha of timber estate on unproductive, degraded and bare lands within the designated production forest; reforestation and rehabilitation of 1.9 million ha of degraded public forest land; and regreening of 4.9 million ha of critical private land (Mok 1990). A substantial area of the convertible production forest is expected to be replaced by industrial agricultural crops, such as rubber and oil palm, which are essentially tree crops with protective functions.

Production Forests in Malaysia

Malaysia seems to be the only other country in the Asia-Pacific region, along with Indonesia, to have an explicit and clearly defined Permanent Forest Estate comprising protection and production forests. It was established in recognition of the crucial role of forests in the production of timber and non-wood forest products; the protection of soil, water and environmental resources; and the conservation of wildlife and biological diversity. The Permanent Forest Estate covers a total area of 12.73 million ha or 38.7 per cent of the total land area of which approximately 9.99 million ha or 78.5 per cent have been identified as production forests. Of the total area of production forests, 3.90 million ha or 39.0 per cent are still virgin while 6.09 million ha or 61.0 per cent have been harvested selectively and are regenerating. The Permanent Forest Estate of Malaysia is distributed roughly equally in Peninsular Malaysia, Sabah and Sarawak at 4.74, 3.35 and 4.64 million ha respectively of which 2.84, 3.0 and 4.15 million ha respectively are production forests. Sarawak plans to increase its Permanent Forest Estate to above 70 per cent of the forested land area by the 1990s. Of the total area of production forests in Peninsular Malaysia, only 0.44 million ha or 15.5 per cent are still virgin while in Sabah and Sarawak, the corresponding areas are 0.75 million ha or 25.0 per cent and 2.71 million ha or 65.3 per cent respectively. The total growing stock in the Permanent Forest Estate of Malaysia has been estimated for all trees 10 cm dbh and above at 2445 million m^3 of which 701 million m^3 are considered to be merchantable volume from trees having 45 cm dbh and above. A substantial portion of the remaining forested land is conversion forests which yielded more than half of the total industrial roundwood production in 1989.

Besides managing the natural forests for sustained yield, environmental protection and biological diversity conservation, Malaysia has embarked on a programme to establish commercial forest plantations. During the 1950s, 779ha of teak were planted in the northern states of Peninsular Malaysia while 5558ha of fast-growing tropical pines were planted during the 1960s and 1970s. The Compensatory Plantation Programme in Peninsular Malaysia was launched in 1982 to establish 188,200ha of fast-growing hardwood timber species over a 15 year rotation of which 36,874ha have been planted by the end of 1989. Sabah plans to establish 250,000ha of fast-growing forest plantations of which 50,306ha have been planted. Sarawak has planted about 4000ha and plans to an additional area of 20,000ha by the year 2000. Apart from planted timber trees, Malaysia has also established more than 25,000ha of rattan plantations, especially in Peninsular Malaysia and Sabah, and small plantations of traditional wild fruit tree species along the fringe of the forested lands which hopefully will be an effective buffer zone to protect the Permanent Forest Estate and contribute towards biological diversity conservation.

Although Malaysia has yet to formally adopt a national land-use policy, development of its land and forest resources has been carried out systematically based on land capability studies and in accordance with sound land-use principles and the concept of sustainable development. Furthermore, most of the forested land converted for agricultural development has been planted with industrial tree crops, such as rubber, oil palm and cocoa, which are structurally similar to closed timber tree plantations and many temperate forests and which effectively maintain the country's vegetative cover at more than 70 per cent of the total land area.

In recent years, Malaysia has developed rapidly. Conflicts between development activity and environmental quality have raised public awareness of the importance of the conservation of nature. In response to public concern, an urban forestry programme has been launched which involves the planting of trees not only to beautify the urban environment but also to improve the habitats for wildlife.

Production Forests in the Philippines

In the Philippines, forested lands are classified generally into timber land, national parks, watershed and wildlife reserves, military and civil reserves, fish ponds, mangrove, brush lands, and alienable and disposable lands. The classification of the country's forested land indicates a continuous change of land use, due mainly to increasing population pressure, which seems to imply that the Philippines does not have a permanent forest estate or permanent production forests. However, the recently completed National Forest Inventory estimated that, as of 1990, the extent of forests under old growth dipterocarp, second growth, and pines, which are essentially the country's major production forests, is 984,100ha or 3.3 per cent of the total land area, 3,455,800ha or 11.5 per cent, and 238,300ha or 0.8 per cent respectively. Forest resource depletion in the Philippines, which ranks amongst the highest in the world, is due mainly to inequitable land distribution. As a consequence, the country has substantial areas of largely degraded and deforested lands which are used for grazing and slash-and-burn subsistence agriculture. Extensively used lands comprise brush land covering 2,459,100ha or 8.2 per cent of the total land area, grass land covering 1,542,900ha or 5.1 per cent, and other uses covering 6,549,800ha or 22.0 per cent.

The effects of rapid deforestation and forest degradation in the Philippines on forest resources and biological diversity have been exacerbated by the highly unsatisfactory rates of reforestation and forest rehabilitation. Many Timber License Agreement holders who are responsible for reforestation failed to fulfil their obligations while the country's reforestation projects covering about a million ha only achieved 27.2 per cent of their target because of financial and other constraints. The poor performance prompted the Philippines to introduce new reforestation schemes which include Contract Reforestation (which involves NGOs, communities and families in the establishment, development, maintenance and protection of forest plantations), Commercial and Industrial Tree Plantations (a private sector initiative which has been unsuccessful due mainly to a lack of financial capital), and Integrated Social Forestry Programmes (which combine previous initiatives such as Forest Occupancy Management, Communal Tree Farms, and the Family Approach to Reforestation into a cohesive programme to designed to help overcome the problems caused by shifting cultivation) (Mok 1990).

A Master Plan for Forestry Development in the Philippines has been completed which accords over-riding priority to identifying and developing projects for a 25 year strategy for the restoration of Philippines forests. Effective implementation of the Master Plan with emphasis on sustainable management of the production forests; reforestation of deforested and degraded sites, especially in critical watersheds and environmentally sensitive areas; and forest protection; will contribute decisively to improved protection of the environment and conservation of forest resources and biological diversity. The proposed ban on logging in selected regions of the country should enhance the prospects for forest resources and biological diversity conservation, if it can be implemented effectively.

Production Forests in Thailand

The natural forest resources of Thailand, apart from the teak forest of the north, are poorly known or understood. There is no recent national forest inventory. The latest data from satellite imagery, taken in 1988, indicates that only 14.38 million ha[1] or 28.0 per cent of the total land area is forested. Of the total, Tropical Evergreen Forest and Mixed Deciduous Forest, which are the major productive forest types, cover 6.23 million ha or 43.3 per cent and 3.12 million ha or 21.7 per cent respectively. Dry Dipterocarp Forest, which is relatively poor, covers a substantial area of 4.50 million ha or 31.3 per cent of the total forest area. The significant increase in forest resource depletion in Thailand in recent years has been exacerbated by the rapid socio-economic development which generated considerable conflict in land-use and inequitable distribution of land. More than a million families depend on forest land for their subsistence agriculture and each family is estimated to use 7.4ha annually of which a third is from the virgin forest. As a consequence of the increasing rates of deforestation and forest degradation, which were greatly facilitated by logging activities and recent natural disasters associated with forest clearance, a logging ban was imposed in 1989 (Mok 1990).

Thailand's latest policy objective is to achieve 40 per cent forest cover in the country, of which 25 per cent would be under production forests. The Thai government intends to achieve this objective by intensifying reforestation and forest rehabilitation. Presently, reforestation activities are undertaken by Forest Villages and timber concessions as well as the public sector for watershed protection and rehabilitation of degraded forests. By 1985, 540,000ha had been reforested and an annual reforestation rate of 48,000ha had been proposed under the Sixth National Development Plan. However, current reforestation efforts by the public sector, which fluctuated between 50,000ha and 30,000ha per year depending on available budget, are grossly inadequate and the situation is not expected to improve significantly as reforestation programmes tend to be accorded low priority. Nevertheless, the ongoing programmes plan to establish between 20,000ha and 24,000ha annually. As an alternative, the private and corporate sectors are being provided with very attractive incentives to invest in industrial forest plantations, which include the lease of degraded forest land at sub-economic rents. The strategy is aimed at accelerating the achievement of the policy objective of maintaining 25 per cent of the land area under production forests which means that Thailand will satisfy virtually all its future timber requirements from plantation forests. A Master Plan for Forestry Development is being prepared for Thailand which will no doubt focus on the achievement of this policy objective.

Production Forests in Papua New Guinea

Papua New Guinea does not have a permanent forest estate or permanent production forests as more than 95 per cent of the land is held under customary ownership. Nevertheless, forests still covered 36,179,000ha or 77 per cent of the land area in 1985 as the country has a relatively low deforestation rate due to its low population density of only 8 people per sq km, which is amongst the lowest in the world. Naturally broadleaved forests and coniferous forests covered 35,623,000ha and 520,000ha[2] respectively while forest plantations covered only 36,000ha. The extent of the forest resources, their ecological characteristics, regenerative capacity and likely yields are all very poorly known. The major production forest types appear to be lowland rain forest, which is the predominant vegetation type, and lower montane forest, where hoop pine is

---

1   14.31 million ha according to unpublished data provided by the Royal Thai Forestry Department.

2   35,200,000ha total forest area according to Saulei (1990)

common. According to an analysis of maps produced from aerial photographs and radar imagery of 1971 over the mainland of Papua New Guinea, lowland rain forest covered 11,061,500 ha or 33.7 per cent while lower montane forest covered 7,618,000ha or 23.3 per cent. In 1974, based on aerial photographs from 1944-5 and the early 1960s, FAO estimated the total potentially utilisable forest area at 15 million ha. As up to 40 per cent of the area is inaccessible or unloggable due to topographic and associated factors, only 6 million ha are likely to be productive forests (FAO 1989).

Production Forests in Nepal

Nepal does not have a permanent forest estate or permanent production forests, apart from the protected areas, and it is unlikely to be able to establish any in the natural forest as pressure on the remaining forests is expected to continue or even increase as a consequence of the growing population and its demand for fuelwood and timber, tree-leaf fodder, and farm land. In Nepal, forested land with at least 10 per cent crown cover has been estimated at 5,518,000ha or 37.4 per cent of the land area. However, the remaining accessible forests have been degraded to such an extent that they now have poor prospects for regeneration. Realising the urgent need for concerted action and a comprehensive framework for the development of the forestry sector, particularly the restoration of the country's forest cover, Nepal has completed a Master Plan for the Forestry Sector which is currently being implemented. The plan includes six primary development programmes including a Community and Private Forestry Programme and a National and Leasehold Forestry Programme. The former is aimed at increasing the supply of forest products by improvement in productivity and accessibility through the management and enrichment of natural forests as community forests; establishment of community forest plantations; and provision of seedlings for planting on farms. The Leasehold Forestry Programme is aimed at strengthening the national forest estate to increase production to meet the needs of wood-based industries and the urban centres through the direct management by the government and the lease of forest land to individuals, private firms, or other bodies, for the commercial production of wood.

Production Forests in India

India has a long history of forest administration and has been a leader in the establishment of forest reserves and the preparation of working plans for forest management based on the principles of sustained yield. The maintenance of a permanent forest estate or permanent production forests is uncertain and will be difficult in the long term as the average annual rate of deforestation in the 1980s has been estimated at about 1.5 million ha. The bulk of the roundwood in India is used for fuelwood. Fuelwood harvesting was estimated at 226.3 million cubic meters[1] or more than 90 per cent of total wood utilisation during the 1985-87 period. The prospects for maintaining the permanent forest estate or permanent production forests, especially for timber production, would seem to be quite poor unless alternative sources of non-wood energy are available.

---

1   240 million m^3 per year according to Collins (1991)

5. Forestry in Biodiversity Conservation

The discipline of forestry is literally as old as the hills and has been praticised traditionally but informally in the Asia-Pacific region from the time when the people began to use the natural forests to satisfy their basic needs. The objectives of forest management then were simple and obvious as the needs were few and the abundant forest resources could easily meet the demands of the relatively small number of people without any undue stress. The requirements of the people were mainly in the form of non-wood forest products, such as food and medicine from plants and animals, fuelwood, and materials for shelter. These items could be harvested with minimal damage to the forests. The people learned to live in harmony with nature. The forests served their spiritual, cultural and recreational needs. Sustainable management was achieved by prudent and balanced use of the forests with minimal impacts on the forest environment, the forest resources, the forest ecosystem, or biological diversity.

Traditional forestry in the Asia-Pacific region was subjugated by conventional forestry with the arrival of the Europeans and the discovery of the region's rich forest resources. Forestry became "the science, the art and the practice of managing and using for human benefit the natural resources that occur on and in association with the forest lands" and "more specifically, the husbandry of tree crops, or the profitable exploitation of the resources intrinsic to forest land". Human benefit and the profit motive assumed predominance and the forest exploiters soon realised that logging is more profitable than the extraction of non-wood forest products. Large-scale mechanisation and the removals of maximum volumes were justified on the grounds of full and efficient utilisation, economy of scale, and maximum socio-economic benefits. However, intensive exploitation, especially of tropical forests, inevitably brought about rapid resource depletion, excessive resource wastage, and above all, adverse impacts on the forest environment, forest resources, and biological diversity. The damage was aggravated by the systematic application of silviculture, often without the necessary information on the residual stand or its ecological characteristics, to reproduce idealistic uniform and highly productive forests – just like those in Europe – seemingly without realising that the complex forest ecosystems in the Asia-Pacific region, particularly those located in the tropical countries with intense population pressures, harsh climatic conditions, and fragile environments, are not amenable to such extreme treatments. Conventional forestry has been alleged to have contributed not only to forest degradation and deforestation but more importantly to inequitable distribution of wealth, deprivation of the indigenous peoples' rights, loss of biodiversity, and environmental problems, notably the greenhouse effects and global warming.

In recent years, the more people-oriented concept of "New Forestry" has been proposed to supplant "Conventional Forestry". It is claimed that New Forestry will rectify the defects and failures of the latter and in the process redress social injustice, resolve the problems of environmental degradation, and ensure the security and integrity of the forest resource base. The approaches to New Forestry are many and varied, given the highly diverse Asia-Pacific region, and have been intensely debated and aggressively promoted, especially by the academics and the non-governmental organisations. When referring to New Forestry natural resource managers refer to the terms such as "Community Forestry," "Social Forestry," "Agroforestry," and more recently, "Multiple-use Forestry" and "Extractive Reserves". The objective is the involvement of the local community in forest management and development, and this, it is claimed, will overcome all the weaknesses of Conventional Forestry and result in sustainable forest management and biological diversity conservation. The rapid proliferation of approaches has, however, generated more confusion than enlightenment which could lead to a further loss of biological diversity, without anyone realising it! In the pursuit of idealistic goals, the proponents of New Forestry seem to have forgotten the realities of life in developing countries and that, in the complex and inter-related natural and cultural environments of the Asia-Pacific region, only a comprehensive and holistic solution based on pragmatic consideration will achieve the desired results.

The International Tropical Timber Organization is committed to support the practice of sustainable forestry. It is based on the belief that successful sustained yield management of the forests will only be achieved if effective control is exercised over the following:

1)   Protection of the forest.

2)   Realistic assessment of annual cut.

3)   Orderly arrangement and demarcation of annual coupes.

4)   Pre-felling inventory and allocation of silvicultural system.

5)   Marking of trees for retention or for felling.

6)   Exploitation of coupe to acceptable damage limits.

7)   Post-felling inventory.

8)   Check of coupe to acceptable damage limits.

9)   Silvicultural treatment of relict stand if necessary.

10)   Continuous forest inventory.

11)   Maintenance of main roads, control of erosion on spurs and skid trails.

While most of the conditions stipulated for sustainable forestry are undoubtedly necessary, they are insufficient to ensure sustainable forest management and will not contribute significantly to the conservation of biological diversity. Many countries in the Asia-Pacific region have adopted the necessary practices but most of them have not been successful in achieving sustainable forest management or biological diversity due primarily to a lack of awareness of alternative strategies and options for forest management and development; insufficient information on the forest and biological resources and their responses to interventions; a lack of appropriate technologies and techniques to assess and monitor the forest and biological resources; inadequate management expertise and operational skills; and insufficient human and financial resources.

"Conventional Forestry" failed to achieve sustainable forest management and biological diversity conservation because its Eurocentric forestry philosophies, concepts, practices and goals are incompatible with the very different political, social, economic and cultural conditions prevailing in the Asia-Pacific region and the highly complex environmental and ecological conditions inherent in the forests of the region. While "New Forestry" could mobilise the peoples' support to protect and possibly enlarge the forest resource base, it is unlikely to be able to contribute significantly to the sustainable management of the remaining natural forest resources, particularly the extensive areas of tropical forests in the hinterland, and will not contribute to biological diversity conservation, except indirectly by relieving the pressure on the natural forests. Even "sustainable forestry" may not contribute effectively to biological diversity conservation if its goals are oriented purely towards the sustainable management of timber production without due priority being accorded to the quantification, evaluation and maintenance of biological diversity values.

The principal measures required to achieve biodiversity conservation in the production forests must include pragmatic objectives and technically sound forestry practices over and above the security of tenure, efficient administration, and effective protection of the forests. Conservation of forests should be based on integrated forest resources assessment and realistic forest management and biodiversity conservation goals, strategies, and options. Emphasis should be placed on preventive or proactive measures not only for sustainable management of the forest resources but also for the conservation of biodiversity. The conservation approach is logical and will be effective as biological diversity is an integral part of the forest ecosystem. A conservation strategy for the sustainable management of the natural forest resources and effective conservation of biodiversity requires the following policy directions:

1) Manage and utilise the forest resources for optimal benefits based on the inherent capability of the forest.

2) Manage the utilisation of the forest resources based on comprehensive forest land use and management plans.

3) Determine potential yield on the basis of systematic and in-depth appraisals of the forest resource base, its growth potential, and other relevant factors.

4) Regulate log flows based on a careful balance of supply and demand as well as maximum utilisation prospects and constraints.

5) Harvest the forest resources conservatively by selective felling and retention of adequate natural regeneration, consistent with economical harvesting, to ensure the sustainability of the forest resource base.

6) Apply optimal forest management regimes developed on the basis of information generated by systematic integrated research.

The conservation approach to forest management is based on the proven concept of management by objectives and endeavors to maximise the following forest resource and environmental management goals:

1) Efficient utilisation of the productivity of the site.

2) Conservation of the genetic and other non-wood natural resources.

3) Maintenance of environmental stability and quality, particularly in sensitive watersheds, as well as water supply and quality.

4) Economic harvesting and utilisation of the crop to be removed.

5) Retention of a viable stand for regeneration.

The effective implementation of the conservation approach requires careful pre-felling and post-felling assessments of the forest resources and the application of appropriate operational guidelines which ensure that the management practices are environmentally, ecologically and economically sound and sustainable. The conceptual sequence of operations for the "conservation" approach to sustainable forest management is as follows:

Year	Operation
N-2 to N-1	Pre-felling integrated forest inventory Determination of optimal forest management regime or option
N-1 to N	Tree marking for felling or retention, if necessary Climber cutting, if necessary
N	Directional felling of prescribed trees
N+1	Post-felling integrated residual inventory Determination of optimal silvicultural regime or option Silvicultural treatment, if necessary

The conservation approach to forest management recognises the political, social, economic, environmental and ecological constraints prevalent in most of the developing countries in the Asia-Pacific region. The approach is consequently based largely on technical, technological, managerial and operational considerations. This realistic and pragmatic approach could be put into practice resulting in the following beneficial effects which:

1) conserve forest resources and biodiversity;

2) ensure maintenance of the forest resource base;

3) minimise reinvestment for reforestation;

4) preserve environmental stability and quality;

5) reduce excessive damage to the forest ecosystem and loss of biodiversity;

6) reduce excessive wastage of harvested resources; and

7) induce optimal utilisation of resources.

The conservation and sustainable management and development of the locally valuable and globally vital forest resources of the Asia-Pacific region can best be achieved by the implementation of "Conservation Forestry" based on the integration of Eastern wisdom and philosophy with Western expertise and technologies. The necessary expertise, particularly in policy and economic analysis, must be acquired and appropriate facilities established within the Asia-Pacific region as a matter of priority to enable the countries in the region to formulate realistic strategies and options for sustainable forest management and biodiversity conservation. There is also an urgent need to upgrade essential technological, managerial and operation skills, not only to intensify integrated studies in forest management and operations to develop the appropriate tools and generate the relevant information but, more importantly, to apply the appropriate technologies, expertise and skills in the sustainable management of the forest resources and the conservation of biodiversity. The most urgently required technologies and skills in most of the countries of the Asia-Pacific region are efficient and cost-effective integrated forest inventory systems, change detection and monitoring systems, database management systems, operations management and decision support systems, and non-wood forest resources assessment and evaluation systems.

In spite of the increasing pressure for change, management of the permanent forest estate in most countries of the Asia-Pacific region is still very strongly entrenched in the concepts and practices of Conventional Forestry introduced from Europe. Although most of the countries have rational and sound national forestry policies, legislation and administrations, which provide the basis for forest conservation, management and development as well as the development of forest industries, implementation and control tend to be weak due to inadequate and insufficient human and financial resources and other constraints. Many of the countries also have separate conservation policies, legislation and administrations with provisions overlapping those for forestry. This tends to create areas of conflict. As a consequence, few forest management and development plans have specific or priority provisions for biodiversity conservation. This could result in serious losses of biological resources. Fortunately, the complexity of the natural forests, particularly tropical forests, in the Asia-Pacific region and the selective nature of the timber markets, have compelled most countries to adopt selective cutting systems in recent years which has indirectly reduced the adverse impacts on the biological resources. Nevertheless, there is an urgent need for rationalisation and harmonisation of policies, strategies and programmes for forest resource management and development and the conservation of biological diversity. An equally urgent need is the acquisition of information on biodiversity within the production forests and its response to timber harvesting and other associated activities.

Indonesia praticises both clear cutting and selective cutting in the management of the country's natural forest resources. The Clear Cutting System with artificial regeneration is widely used in timber estate development, particularly in teak, pine and other forest plantation management on Java and in the conversion of poorly stocked, unproductive forests into industrial forest plantations. The Indonesian Selective System is designed for use in Dipterocarp forests having a normal stem-diameter frequency distribution and where there is sufficient regeneration for future forest development. Selection of trees to be harvested is generally based on species and diameter which may vary according to local forest and market conditions. General conditions and broad guidelines for sustainable management and development of the forest resources have been formulated but are often not enforced effectively due to inadequate human and financial resources and other constraints. Enrichment planting has been made mandatory recently to ensure adequate regeneration. If conditions imposed to achieve sustainable management are not observed strictly and operations are not supervised carefully, however, the relatively high intensity of removals and heavy equipment used in logging will have detrimental effects on biodiversity conservation. On the other hand, the generally lax silvicultural requirements could be a blessing as the effects of logging on the biological resources will not be aggravated by premature re-logging of the forest.

A forest concession system was introduced in Indonesia in 1967, to achieve fuller utilisation of the mixed dipterocarp forests. The system provides for the allocation of 20-year concessions over specified areas of forest land for harvesting, regeneration and maintenance as well as for processing and marketing the forest products in accordance with plans agreed upon between the concessionaires and the Ministry of Forestry. Under the system, concessionaires are required to praticise sustainable forest management and development, based on sound conservation and sustained yield principles. They must prepare three types of working plans: 20-year overall concession work plans, five-year working plans, and annual work plans. Concessions which expired recently were required to prepare and submit new working plans supported by comprehensive forestry inventory information and detailed proposals for consideration and approval. More stringent conditions are expected to be imposed in response to increasing global pressures for sustainable forest management. This should contribute positively to the conservation of biodiversity.

A major problem in biodiversity conservation in the production forests of Indonesia is the tendency for logged-over forests to be encroached upon by the local people for shifting agriculture. Protection of these forests is extremely difficult due to the large area involved, the easy access provided by logging roads, the limited numbers of forestry staff, shortages of funds, logistical problems, and other constraints. Consequently, extension and education programmes for land rehabilitation and natural resources conservation have been intensified. A Concession Guided Village pilot project has been launched in Kalimantan to assist the settlers to cultivate the land permanently and to involve them in the establishment of forest tree plantations as a supplementary source of income. The programme is expected to be expanded rapidly and significantly by the relevant agencies to ensure that the production forests play a more positive role in the conservation of biodiversity.

In Malaysia, the natural forests, especially the lowland Dipterocarp forests, have traditionally been managed under the Malayan Uniform System (MUS). The System involves the removal of the mature crop in one single felling of all trees down to 45cm dbh and the release of the selected natural regeneration of varying ages which are mainly the light demanding medium and light hardwood species. The felling operation is followed by poison-girdling of defective relics and non-commercial species down to a minimum 15cm dbh. Approximately 3-5 years after felling, a post-felling forest inventory is carried out to determine the presence and status of regeneration on the ground and subsequently to determine suitable silvicultural treatments. As the MUS originally relied on the seedlings and saplings to form the next crop, the silvicultural treatments were prescribed to favour these groups, often at the expense of bigger trees and advanced growth. This led to a much heavier poison-girdling of trees than was necessary, and in some cases, too drastic opening of the canopy which proved to be detrimental to sustainable forest management and biodiversity conservation. The shift from seedlings and saplings to advanced growth for regeneration in recent years led to a more discriminating use of poison-girdling and a more conservation-oriented approach to silvicultural treatments, which reduced the loss of biodiversity significantly.

Recently, Peninsular Malaysia formulated and implemented the Selective Management System (SMS) which is based on the concept of Conservation Forestry. The following strategies have been adopted to achieve sustainable forest management and development:

1) Forested land scheduled for conversion to agriculture will be released in a planned and controlled manner to ensure a regulated and reliable flow of logs and maximum utilisation.

2) Forested land not due for conversion will be designated as permanent forests to be managed as a renewable resource for maximum commercial returns compatible with the maintenance of environmental quality and biodiversity.

3) A systematic pre-felling forest inventory of not less than 10 per cent intensity will be conducted to determine the optimal forest management and silvicultural systems to be applied.

4) Post-felling forest inventories will be conducted to assess the status and composition of the regeneration, and to determine appropriate silvicultural treatments.

5) Comprehensive soil surveys will be conducted to locate suitable sites and to determine optimal species for forest plantation establishment.

6) Multiple-use forests areas will be determined and reserved for recreation and conservation of soil, water, flora and fauna.

The SMS was conceived to reconcile the often conflicting objectives of sustainable management, economic timber harvesting, and the costs of reforestation. It requires the selection of optimal forest management options, which are equitable to both the logger and the forest owner, based on data from the pre-felling forest inventory as well as from growth and yield analysis, forest operations, and market studies. Consequently, continuous forest resources monitoring, integrated studies in forest management and operations, and market intelligence are essential adjuncts for the effective implementation of the SMS. The SMS was also designed to allow for flexibility in timber harvesting which would enable the forest manager to take advantage of the vagaries of the timber markets, safeguard the fragile environment and ecosystems, and maximise the conservation of natural forest resources and biodiversity. In Peninsular Malaysia, however, the following prescription is generally followed:

1) The cutting limit prescribed for the group of dipterocarp species should not be less than 50 cm dbh, except for *Neobalanocarpus heimii* (Chengal) where the cutting limit prescribed should not be less than 60 cm dbh.

2) The cutting limit prescribed for the group of non-dipterocarp species should not be less than 45 cm dbh.

3) The residual stocking should have at least 32 sound commercial trees per ha for diameter class 30-45 cm or its equivalence.

4) The difference in the cutting limits prescribed between the dipterocarp species and that of the non-dipterocarp species should be at least 5cm.

5) The percentage of dipterocarp species in the residual stand for trees having 30 cm dbh and above should not be less than that in the original stand.

Neither the MUS nor the SMS provides directly and specifically for biological diversity conservation. Nevertheless, such provisions are inherent in the National Forestry Act of Malaysia which requires the designation of forest areas as wildlife sanctuaries, virgin jungle reserves, and amenity forests for various purposes including biodiversity conservation. Similar provisions are found in the respective State Forest Enactments and Rules which are often augmented by specific conditions in logging licenses. Protection of the forest environment and conservation of forest resources and biological diversity are also achieved by operational guidelines such as those for the construction and maintenance of forest roads and for forest harvesting.

In the Philippines, management of the natural production forest is anchored on the Selection Logging System which was developed in the late 1950s. The basic idea of the system is to leave an adequate number of undamaged trees after logging to permit a second cut after the end of the cutting cycle. The system was designed to make possible another economic cut after a cutting cycle of 30, 35, 40 or 45 years, depending on the growth and climatological conditions in the area. The system begins with a tree marking phase which consists of the laying out "set-up" boundaries, pre-marking sampling, computation of the marking goal, and tree marking. The logging set-ups serve as the smallest management unit in the field for administrative and record-keeping purposes. Pre-marking sampling provides information on the species and diameter distribution in the set-up, and the results of sampling are used to determine the number of trees that must be left in the set-up as residuals (the "marking goal"). Based on the results of sampling, 79 per cent of commercial trees between 20 and 60 cm dbh and 40 per cent of those in the 70 cm dbh class are marked as residuals. After felling, an inventory of residuals is conducted to (a) determine the extent and nature of damage caused by the logging operations so that necessary corrective measures can be taken; (b) to appraise the residual growing stock for yield projections; and (c) to determine the fines to be paid by the timber licensee for the damaged residuals. Timber stand improvement is carried out five to ten years after logging. The operation consists of selecting the potential crop trees, and the removal or killing of unwanted species, deformed trees, and climbers.

Each forest working unit, or area under a Timber License Agreement, has a prescribed annual allowable cut, determined on the basis of prior knowledge of the volume and size-distribution of commercial species in the area, and using growth-prediction models applicable to the climatological region where the working unit is located. This system is being reviewed and is expected to be phased out eventually. The discussions centre on the validity of the system *vis-à-vis* the sustained yield objective, the tendency of the system to favour traditional species, and on the question of whether eliminating the so-called unwanted species and reducing the species diversity of the forest is ecologically sound. In addition, it has been noted that the original growth estimates upon which the management regime depends so much, were mostly derived from the lowland dipterocarp forests which were better stocked than the present old-growth forests. Presently, the Timber Production Sharing Agreement (TPSA) model is being tried out in several pilot areas. The results of these pilot tests shall be used as a basis for setting up a comprehensive system to replace the old Timber License Agreement (TLA) system. In the meantime, the TLAs which are still in effect shall continue to operate as such, on condition that the TLA holders comply with the more rigid prescriptions for forest management which are being promulgated.

The TPSA mode is a stumpage valuation scheme. The private investor is given a specific area and volume to log; the investor shoulders all costs of operation – from road building to log transport. The after-costs proceeds of timber sale is determined, using current log prices, which is then shared by the investor and the government based on a pre-determined sharing allocation. In its present form, the TPSA implies a greater responsibility for the government in the management and protection of the forest resources. Post-harvest operations, such as timber stand improvement, enrichment planting, reforestation and forest protection, become government's direct responsibility, in contrast to the TLA system where the timber licensee was directly responsible for these operations. In the TPSA system, the investor's tenure is short, thus, long-term development and management work necessarily becomes government's direct responsibility.

The Philippines has completed a Master Plan for Forestry Development which will identify and develop projects as components of a 25 year strategy for the restoration of Philippine forests. Some of the issues to be considered in the forestry sector review mission which will identify suitable projects for incorporation in the Forestry Master Plan include (Anon. 1989):

1)   Preparation of short, medium and long-term projections for timber and fuelwood demand, based on domestic consumption of industrial round wood, wood products, wood pulp and fuelwood, and estimated export demand for wood-based products.

2)   Participation of local communities in reforestation and other forest management schemes, including an assessment of the characteristics of the local communities now involved in traditional agriculture and their attitudes towards established livelihood patterns, forest management and logging.

3)   Examination of the role of the private sector in Philippine forestry, including the prospects for promoting commercial tree plantations on privately owned lands.

4)   The development of proposals for the establishment of new mangrove plantations in non-productive mangrove and mud flat areas.

5)   Formulation of a long-term strategy for the cultivation of minor forest products (e.g. medicinal plants) of known potential.

6)   Identification of critically degraded watershed areas for rehabilitation and protection.

In Thailand, all classified forest lands are state owned and are managed by the Royal Forest Department of the Ministry of Agriculture and Cooperatives. According to the National Forestry Policy adopted in 1985, the existing forested area is expected to be increased to 40 per cent of the total land area by the rehabilitation of degraded forest lands and wastelands and by establishing large-scale forest plantations. Of the 40 per cent of forest lands, 25 per cent will be reserved for productive and 15 per cent for protective purposes. The government's target of 40 per cent forest cover to be achieved through plantations has been considered to be quite unrealistic as there may not be sufficient land with the potential to justify the heavy investments required to establish and maintain industrial plantations. Furthermore, there may not be markets for the vast amount of low-grade timber that would be produced (WCMC 1990).

Prior to the recent logging ban, forest management in Thailand was confined mainly to logging control and the establishment of small-scale plantations. At the time, most of the forests classified for production were under long-term concessions. The production forests of Thailand were managed under a selection cutting system with the concession area divided into 30 annual coupes of almost equal size. All trees above the allowable cutting limits were selected and marked one year prior to timber cutting and approximately 30 per cent of these selected trees were reserved for the next felling cycle while the rest could be exploited. The allowable cutting limits of each species was designated by Forest Acts and varies from 20cm to 150cm dbh. Timber cutting was operated by annual coupe and the harvested forest was usually left to natural regeneration to produce another cut after 30 years. The concessionaire was responsible for reforestation and fire protection, but compliance with these conditions tended to be unsatisfactory. In recent years, encroachment into the logged-over forest for both shifting and permanent agriculture has increased significantly due to the easy access provided by the logging roads. The ban on logging of natural forests imposed in 1989 should check further encroachment and stop further forest degradation and deforestation in Thailand and thus contribute positively to the effective conservation of the country's forest resources and biodiversity. The Master Plan currently being prepared offers another opportunity for improving the prospects for forest resources and biodiversity conservation.

In Papua New Guinea, large-scale commercial logging is a relatively recent phenomenon as the country is largely inaccessible while the extent of the forest, its ecological characteristics, regenerative capacity and likely yields are all very poorly understood. Systematic forest management is further confounded by the fact that virtually all the natural forests are on lands held under customary ownership and are, therefore, regarded as private property. Nevertheless, the country's constitution provides "for Papua New Guinea's natural resources to be conserved and used for the collective benefit of us all and to be replenished for the benefit of future generations". However, there seems to be no formal application of a sustained yield management system, as the sustained yield objective is not being met. The rate of timber cutting greatly exceeds the rate at which it is being replaced (FAO 1989). Felling is generally selective, removing only the more highly valued species, after which the forests are left to recover without silvicultural management. The 1989 TFAP mission detected lowering of diversity and poor regrowth of valuable species and concluded that the existing management system has reduced the natural growing stock by significant amounts.

Overall responsibility for the conservation and management of the forest resources of Papua New Guinea is defined in the Forestry Act. It empowers the minister for forests to acquire the rights to harvest timber from willing customary landowners and arrange for logging by private industrial companies. Alternatively, private landowners can deal directly with the developers under the Forestry and Private Dealings Act, subject to the approval of the minister in order to safeguard the owners' and the nation's interests. A large number of concessions have been granted but sustainable management is uncertain as a major problem in achieving sustainable yields is that little is known about rates of natural and man-made deforestation before and after logging, and about the proportion of concession areas physically accessible. On the other hand, prospects for biodiversity conservation may be brighter in the absence of silvicultural management, and especially with the environmental safeguards included in the concession agreements which prohibit logging within 20m of permanent water courses, 50m of major rivers, and on land with gradients above 30 degrees.

In Nepal, more than 80 per cent of the country is covered by rugged hills and mountains which are not amenable to timber production, and the remaining accessible forests consist mostly of stands with poor prospects for regeneration. Forest management in Nepal is concerned mostly with protection and rehabilitation of the highly degraded forests and deforested lands. The role of forestry, especially production forestry, under such circumstances, seems to be very limited and insignificant. This has been acknowledged in the Master Plan for the Forestry Sector of Nepal, which has conservation as the central theme. The Plan is based presumably on the logic that conservation of forest resources is a necessary condition for obtaining a sustainable yield from those resources and emphasises that forest conservation is not a separate matter but an integral component of production forestry. Although the Master Plan is conservation oriented, its policy states that "land and forest resources will be managed and utilised on a long-term basis, according to their ecological capability, so as to conserve the forests, soil, water, flora, fauna, and scenic beauty".

The Master Plan for the Forestry Sector of Nepal includes two primary development programmes concerned with production forestry. The first of these, the Community and Private Forestry Programme, was designed to reduce the demand for forest products and to increase the supply. Through improvements in productivity and accessibility, it will focus on the management of natural forests and enrichment planting in degraded forests. Community forest plantations will be established in open and degraded areas and free or subsidised plants will be distributed for planting on farms. The second component of the Master Plan is the Natural and Leasehold Forestry Programme which is intended to be the major instrument for production forestry and is designed to increase timber production by intensifying sustainable management of the natural forests, and where appropriate, by leasing the forest land to the private sector for the commercial production of wood to meet the needs of the wood-based industries and of the urban centres. It is hoped that the increase in sustainable timber production will lessen the pressure on the remaining natural forests which will contribute significantly to biodiversity conservation in Nepal.

In India, about 50 per cent of the total forest area is constituted as reserved forests under the Indian Forest Act. These forests are managed systematically on the basis of regular working plans which regulates felling and prescribe a specific set of treatments under an appropriate silvicultural system. The main components of a silvicultural system applied to natural forests in India are:

(a) harvesting the tree growth that already exists;

(b) regeneration of the felled area; and

(c) tending the regeneration to maturity to achieve pre-determined objectives
    (FAO 1990).

As the cradle of scientific forest management in the tropics, India has evolved a whole complex of silvicultural systems to suit the wide range of climatic and ecosystems in the country. They include:

(a) **Selection Systems** involving the selective removal of valuable species in evergreen and semi-evergreen, moist deciduous forests;

b) **Shelterwood Systems** comprising the Uniform System which involves complete felling spread over the regeneration period in moist deciduous forests where regeneration of valuable species is adequate, and Irregular Shelterwood System which involves felling of all trees above a specified girth spread over the regeneration period in evergreen and moist deciduous forests where regeneration can be established with considerable effort;

(c) **Clear-felling Systems** involving complete removal of standing growth in one operation and natural regeneration in most deciduous forests where natural regeneration is very profuse, or artificial regeneration in moist and dry deciduous forests where natural regeneration is unsatisfactory or where a change in crop composition is desired; and

(d) **Coppice Systems** involving complete felling or the retention of specified numbers of standards in dry deciduous forests to produce fuelwood and small timber.

No reliable data are available on the areas under different silvicultural systems as a whole. However, selection or selection-cum-improvement felling and coppice with standards appear to be the most important forest management systems. Selection felling, which is the most widely adopted silvicultural system in India, is applied in the tropical mixed forests due to:

(a) low proportion of valuable species;

(b) lack of information on end-uses;

(c) poor accessibility; and

(d) environmental protection.

The main objectives of management for evergreen forests under the selection system are:

(a) maintenance of tree cover to protect the soil and to regulate water yield in catchments;

(b) increasing the supply of wood and other products; and

(c) consistency with the above objective realising the maximum revenue.

Although the adoption of low intensity selective felling is primarily determined by economic constraints and is not always an outcome of integrating environmental considerations, it has the indirect benefit of maintaining the prospects for biodiversity conservation. Despite the long history of forestry in India, society has treated natural forests as a freely available resource, and as a consequence, uncontrolled exploitation still continues.

Organisational changes in forest administration have been marginal and have not brought about any significant change from the traditional policing role. As long as such a situation persists, sustainable management of the tropical mixed forests will not be possible. Analysis of management systems indicates the limitations of present approaches. The creation of intensively managed plantations on barren land to meet most of the wood requirements seems to be the only alternative. Indications are that Indian forestry is entering a transition phase during which current practices will be critically examined and a more rational approach to land use, in particular forestry, will have to be pursued (FAO 1989).

6. Towards Effective Conservation

The conservation of natural forest resources and biodiversity is second nature in the Asia-Pacific region where it is generally accepted as a matter of the highest priority, perhaps even as a matter of life and death in some countries. The indigenous peoples know the value of the flora and fauna and use them wisely and prudently for their well-being and welfare. The traditional religions and cultures preach the vital importance of living in harmony with nature, and this has enabled the countries and their peoples to endure and progress. The countries in the Asia-Pacific region recognise the crucial need to achieve sustainable forest management and development and the effective conservation of biodiversity.

Unfortunately, some of the countries in the region, notably in South Asia, are so poverty stricken and have such large and rapidly growing populations that the only recourse left to the people for survival is to clear the forests for subsistence agriculture to feed themselves, to collect fuelwood to cook their food, and for grazing livestock to supplement their meagre diets. In these countries, deforestation, forest degradation and loss of biodiversity were not by choice but occurred as a matter of necessity. This was exacerbated by past mismanagement of the renewable natural forests resulting in rapid resource depletion and environmental degradation. There is little or no hope or scope for major improvements in biodiversity conservation in most of these countries, unless the peoples' dependence on forest lands and products can be reduced by effective socio-economic development outside the forestry sector. In the interim, the only salvation lies in massive land rehabilitation by aggressive reforestation programmes and forceful and sustainable management of the remaining natural forest heritage.

Fortunately, many of the countries in the Asia-Pacific region, notably in South East Asia and Oceania, were well endowed with natural forests but have relatively small populations and low population densities. These countries have been able to exploit their natural forest resources to achieve socio-economic progress and improve the quality of life for their people, apparently without any detrimental effect on the environment and conservation. Through time, however, the adverse impacts of deforestation and forest degradation have become increasingly evident, even though their effects may not be as dramatic, serious or disastrous as the world is being led to believe. The urgent need to improve, so as not to "kill the goose that lays the golden egg", and more importantly, so as to ensure that the people are not threatened or endangered by resource exhaustion, loss of biodiversity, and local environmental degradation, has been recognised and concerted actions to achieve sustainable forest management and development and more effective biodiversity conservation in the production forests have been initiated.

The countries in the Asia-Pacific region have very diverse needs and priorities for improving sustainable forest management and development and achieving more effective forest resource and biodiversity conservation. Most of their needs can only be met by the countries themselves, or with bilateral technical assistance, as the problems and constraints are essentially within their control. Many of the countries in the region have been, and are still, receiving such technical assistance but the achievements have been few and unimpressive. While bilateral technical assistance will continue to be needed and made available, a further proliferation of such technically-assisted projects in the Asia-Pacific region will most likely confuse the issues and compound the problems rather than resolve them. In the final analysis, there is no substitute for quality capability, be it intellectual, managerial, technological, or operational. The countries in the Asia-Pacific region must upgrade the capability of their human resources in the relevant areas necessary to achieve sustainable forest management and development and the effective conservation of forest resources and biodiversity.

In a region of immense diversity, it would be naive to think of a common strategy or general solutions to resolve the problems of deforestation and forest degradation in order to improve the prospects for sustainable forest management and biodiversity conservation. Appropriate development options and action programmes for optimal forest resource and biodiversity conservation will need to be location-specific to be effective and would have to be based on a careful and pragmatic appraisal of the existing resources, socio-economic goals, past trends and future needs, major issues affecting the sector, and relevant development concerns and priorities. Nevertheless, most of the developing countries have similar conditions, needs and priorities, and subscribe to the following general objectives for forestry development (Rao 1990):

1) Sustainable development of a variety of ecosystems (including tropical rain forests) of ecological and economic importance.

2) Minimising ecological degradation and maintaining the productivity of all forest lands through appropriate systems of forest management.

3) Producing forest products on a sustained basis for consumption by the local communities as well as for the overall economic development and export earnings.

4) Generating forest-based income and stable employment opportunities.

As a consequence of the unprecedented and relentless attack on the forest resources during the past few decades, the following seem to have emerged as the priority forestry development concerns in the Asia-Pacific region (Rao 1990):

1)   Managing remaining natural forest areas in a sustainable way.

2)   Involving local communities in the protection and management of natural forest resources.

3)   Identifying critical watersheds and undertaking programmes of conservation.

4)   Finding appropriate solutions to minimise forest degradation by shifting cultivators.

5)   Establishing networks of protected areas to conserve flora and fauna.

6)   Strengthening conservation, collection and storage of germ plasm and tree improvement to ensure productivity gains in reforestation programmes.

7)   Reducing the wastage of wood through improved logging and utilisation.

8)   Utilisation of non-wood products and plantation-grown wood.

9)   Promoting community forestry and agroforestry.

Although a regional strategy and action programme for sustainable forest management and biodiversity conservation in the Asia-Pacific region is an impossible dream, there are distinct possibilities for sub-regional strategies and programmes in areas with similar conditions, problems and aspirations. On the basis of the concerns identified, Rao (1990) proposed the following strategy for future regional forestry programmes for the Asia-Pacific region which centre on three sub-regional projects:

Forestry Development in Continental Asia to address the collective needs of Pakistan, India, Nepal, Bhutan, Bangladesh and China, which are influenced by the Himalayan ecosystem. Among their common preoccupations are appropriate land use; watershed management; upland conservation; interactions in an upland/lowland continuum; agroforestry; community forestry; tree growing by rural population; rehabilitation of degraded forest lands; management of semi-arid and arid lands; etc.

Forestry Development in South East Asia to meet the regional and common aspirations of Myanmar, Thailand, Laos, Vietnam, the Philippines, Malaysia and Indonesia, which are dominated by tropical rain forests. These countries should carefully reassess their forestry and land-use policies to ensure sustainable timber removals in future by bringing unplanned deforestation under control; rehabilitating some 50 million ha of previously logged forests; rigorously implementing sustainable systems of forest management; and undertaking compensatory reforestation schemes.

Forestry Development in the South Pacific to cover the relatively underdeveloped island nations where unique tenurial systems prevail. Among the constraints for forestry development in the South Pacific countries are lack of detailed estimates on the extent of forest resources; lack of the minimum data needed for planning forest management; lack of market intelligence to maximise export earnings or tap new markets through a regional approach; absence of comprehensive planning/programming process linking land use, forest resources, processing, marketing, domestic consumption, socio-economic development; lack of entrepreneurial skills and motivation; and weak institutions.

In the longer term, the problems and constraints of sustainable forest management and development and the associated issues of forest resources depletion and loss of biodiversity in the Asia-Pacific region can only be resolved effectively by formulating and adopting more comprehensive, holistic and realistic forestry development policies, strategies and programmes which accord high priority to the following areas:

1)   **Sustainable management** of the natural forests, especially the tropical rain forests.

2)   **Afforestation of degraded and deforested areas**, especially in critical watersheds and sensitive areas.

3)   **Research and development** to evolve more realistic policy and strategy options for forest management and biodiversity conservation, including the following components:

   (a) to develop appropriate forest management, harvesting and reforestation technologies which are environmentally and ecologically safe;

   (b) to upgrade modern managerial and operation skills;

   (c) to generate relevant and reliable information on forest and biological resources and forest operations; and

   (d) to develop or adapt efficient and cost-effective tools and methodologies for the acquisition, management and dissemination of such information.

4)   **Public education** towards appreciation of the natural forests and awareness-raising amongst forest staff to praticise sustainable forest management and effective biodiversity conservation.

Most countries in the Asia-Pacific region have formulated strategies and programmes for forestry development in general and for forest conservation, management and development in particular, usually with external technical assistance. Many of these strategies and programmes have been found to be ineffective or cannot be implemented, for any of the following reasons: they tend to be idealistic or inappropriate; overly costly and beyond the financial capacity of the countries concerned; require institutional and human resources which are often lacking in developing countries; are politically, socially or economically unacceptable; or are ecologically and environmentally unsound. The social, economic and environmental problems associated with deforestation and forest degradation and the ensuing depletion of forest resources and loss of biodiversity can only be resolved with realistic and pragmatic strategies and programmes for sustainable forest management and biodiversity conservation. The approach based on the concept of Conservation Forestry offers the best prospects for achieving sustainable forest management and development and for the effective conservation of forest resources and biodiversity in the Asia Pacific region, especially in those countries still well endowed with natural forest resources.

The formulation and implementation of realistic strategies and programmes for forestry development in general, and for tropical forest conservation in particular, can only be accomplished successfully after the evolution of a local tropical forestry culture which is compatible with the inherent ecological, environmental, political, social, cultural, and economic conditions. The sustainable management of the natural forests, especially the tropical rain forests, of the Asia-Pacific region require highly competent and motivated forest managers and operators with the relevant expertise and skills; appropriate technologies and methodologies; timely and reliable information; and adequate financial resources. The forestry staff will need to acquire or upgrade their expertise and skills to conduct policy and economic analysis for optimising sustainable forest management and biodiversity conservation; to develop or adapt appropriate technologies and methodologies for environmentally-safe and ecologically-sound forest utilisation and timber harvesting which will ensure sustainable forest management and biodiversity conservation; to conduct cost-effective integrated forest inventories to generate information for integrated resources management; to monitor changes in both the natural and regenerating forest resources; to assess and evaluate the forest and biological resources including non-wood forest products; and to conduct integrated studies in forest management and operations to evaluate the physical and economic impacts of alternative forest management, harvesting and reforestation options on forest resources and biodiversity conservation.

The evolution of a tropical forestry culture, the formulation of realistic strategies and programmes for sustainable forest management and tropical forest conservation, and the effective implementation of projects will require immense technical and financial resources to acquire or upgrade local capacities and capabilities. As most of the countries in the Asia-Pacific region are unlikely to be able to mobilise these resources nationally, they should be encouraged to collaborate and support the establishment of the following with external technical assistance:

1) An **Asia-Pacific Forestry Centre** to conduct policy and economic studies necessary for the evolution of a tropical forestry culture or regional/local forestry culture; formulation and evaluation of strategic options for sustainable forest management and effective biodiversity conservation; and transfer of expertise and skills to national forestry personnel.

2) **Regional and National Centres of Excellent for Sustainable Forest Management and Development** to develop or adapt, package and transfer appropriate technologies, methodologies, and techniques for integrated forest inventory; resource assessment, evaluation and monitoring; timber harvesting and reforestation; information management; and integrated studies in forest management and operations.

3) **Regional Forestry Development Programmes** to provide technical advice, guidance and support and to upgrade the necessary national managerial, technical, technological and operational expertise and skills to ensure sustainable forest management and development and effective biodiversity conservation.

Apart from the core institutions and programmes, the following could be established to provide support for regional and national activities which would contribute indirectly to the conservation of the natural forest resources and biodiversity:

1) An **Asia-Pacific Development Fund** to assist the countries in the region to rehabilitate and reforest degraded and deforested lands, especially in critical watersheds and other sensitive areas.

2) An **Asia-Pacific Forest Industries Development Fund** to support research, development and transfer of appropriate technologies for higher value-added timber-based products and for the production, harvesting, processing, marketing and promotion of non-wood forest products.

3) An **Asia-Pacific Forestry Information Centre** to promote greater awareness and appreciation of the values of forests and to provide technical advice, guidance and support for upgrading national capacities and capabilities.

4) A **Global Forestry Convention** to ensure sustainable forest management and development and biodiversity conservation as an integral component of protection, production and amenity forests.

Bibliography

Bista, R.B. 1990. Forestry sector development: its role in the conservation of biological diversity in Nepal. Unpublished report prepared for the IUCN Forest Conservation Programme workshop on "Realistic Strategies for Tropical Forest Conservation" in Perth, Australia.

Chiew, T.K. 1990. Forest conservation and management practices in Malaysia. Unpublished report prepared for the IUCN Forest Conservation Programme workshop on "Realistic Strategies for Tropical Forest Conservation" in Perth, Australia.

Collins, N.M., Sayer, J.A., and Whitmore, T.C. 1991. *The Conservation Atlas of Tropical Forests: Asia and the Pacific*. Macmillan Press Ltd., London. 256pp.

Ministry of Forests and Soil Conservation. 1988. Master Plan for Forestry Sector – Nepal, Kathmandu.

Mok, S.T. 1989. Natural resource conservation and forest management. Unpublished paper prepared for the International Conference on National Parks and Protected Areas, Kuala Lumpur.

Mok, S.T. 1990. Sustainable management and development of tropical forests in ASEAN. Unpublished paper prepared for the ASEAN Seminar on Management of Tropical Forests for Sustainable Development, Jakarta.

Mok, S.T. 1991. Strategic options for forestry development in Asia-Pacific countries. Unpublished paper prepared for the Regional Conference on Environmental Challenges for Asian Pacific Energy Systems in the 1990s, Kuala Lumpur.

Nuevo, C.C. 1990. Conservation efforts and biodiversity awareness in Philippine forestry. Unpublished report prepared for the IUCN Forest Conservation Programme workshop on "Realistic Strategies for Tropical Forest Conservation" in Perth, Australia.

Rao, Y.S. 1990. Asia-Pacific forestry: regional dimensions. Unpublished paper prepared for the DSE-AIFM Seminar on Integration of Management of Tropical Forest into Regional Development, Kuala Lumpur.

Royal Thai Forestry Department. Data derived from the Annual Report 1990, published in Thai.

Saulei, S.M.. 1990. Untitled. Unpublished report prepared for the IUCN Forest Conservation Programme workshop on "Realistic Strategies for Tropical Forest Conservation" in Perth, Australia.

Sormin, H. 1990. The role of production forests in biological diversity conservation in Indonesia. Unpublished report prepared for the IUCN Forest Conservation Programme workshop on "Realistic Strategies for Tropical Forest Conservation" in Perth, Australia.

United Nations Food and Agriculture Organization (FAO). 1989. *Review of Forest Management Systems of Tropical Asia*. FAO, Rome.

World Conservation Monitoring Centre (WCMC). 1990. *Global Diversity 1992*. World Conservation Monitoring Centre, Cambridge. Unpublished draft.

World Resources Institute (WRI). 1990. *World Resources: 1990-91*. Oxford University Press. Oxford. 383pp.

COUNTRY STUDIES

LATIN AMERICA

BOLIVIA

Based on the work of
Maria Marconi and Ivan Morales

1. Conclusions and Recommendations

The natural resource management situation in Bolivia is unsatisfactory. The state exercises minimal control in both protected areas and production forests. As a result the long-term supply of timber is threatened as is the biological diversity of the country.

In 1986, the Environmental Law (Ley del Medio Ambiente) was first proposed, to coordinate all previous acts regarding the environment and natural resources into a coherent form. A new Project for the General Environmental Law (Proyecto de Ley General del Medio Ambiente) was formulated in 1991, and by the end of the same year, the law was in the process of being approved by National Congress (Congresso Nacional) with a good possibility of being passed (Marconi *in litt.*, 1991).

There is an urgent need to create a coherent and comprehensive legal framework to govern the rational use and protection of forest resources. This will have to be accompanied by institutional strengthening and improved inter-institutional coordination. Generating public support for these measures will require education and awareness campaigns. Parks and forest concessions need to be brought under effective conservation management.

2 Extent, Status and Security of TPAs

The present system of protected areas is comprised of two parts; forest reserves and protected areas (as defined under the 1975 Wildlife and National Parks Law). According to the *Protected Areas of the World: A review of national systems* (WCMC 1992):

> Five categories of protected area are defined in 1975 Law, and four of forest reserves in the 1974 Forestry Law, but in practice, eleven designations are used. The categories not defined in the legislation are fiscal reserve and biological station.

This breaks into four categories of protected areas which correspond to the following IUCN categories:

1) National Park (II), and Wildlife Reserve (IV)

2) Wildlife Refuge (IV)

3) Wildlife Sanctuary (IV)

4) Hunting Reserve[1]

The *1990 UN List of Parks and Protected Areas* estimates that the totally protected area estate includes 6,774,165ha representing 6.2% of national territory and is comprised of 23 management

---

1   Under the current system, there is not an equivalent IUCN category.

units. This includes only areas greater than 1000ha and excludes those whose legal status is uncertain or for which data is lacking. Only three of these units, Amboro and Noel Kempff Mercado National Parks and the Beni Biological Station are receiving any effective protection. Together these three areas represent less than 1% of Bolivia. The majority of the protected areas are not managed. In twelve of these units, government authorities are actually planning exploitative development.

The protected area system was selected without consideration of ecological criteria. New areas have been created with no consideration of existing conservation units, and very little is known about their biogeographical characteristics. Information on the distribution of biological diversity value among different management units is lacking.

All national parks and protected areas in Bolivia are subject to human influence including mining, hunting, pasture conversion, timber harvesting, colonisation and urbanisation.

Management of wildlife and protected areas is the responsibility of the Department of Wildlife and National Parks (DVSPN), under the jurisdiction of the Forestry Development Center (CDF) which, in turn, is a semi-autonomous unit within the Ministry of Agriculture and Farmer Affairs (MACA). The DVSPN lacks a clear policy orientation which would allow for orderly programme development. Frequent changes of directors and lack of funding and qualified personnel have caused bureaucratic weakness. The documentation needed for effective planning, such as maps and biological information, is not available. Coordination with other divisions of MACA, which is needed for integrated land use planning, has not occurred. DVSPN activities have been confined, in large part to the granting of hunting and wildlife export permits. There is little or no communication between the administrative organisations responsible for protected areas.

Conservation laws are not consistent and the categorisation system is confusing. This makes it difficult to evaluate the true extent of Bolivia's totally protected area estate. The selection of protected areas in general has not corresponded to critical sites. Some of the protected areas are the result of preliminary studies. Many of the protected area selection sites suffer from the lack of criteria to identify areas of high biological diversity.

Bolivian law describes approximately one hundred legal measures designed to protect the environment. These are isolated environmental initiatives which are often anachronistic or contradictory. There is no coherent national conservation strategy. Existing laws are often ignored and manpower and funding needed for enforcement are inadequate.

The General Environmental Law (mentioned in section 1) proposed the creation of a Secretariat of the Environment (Secretariat del Medio Ambiente) with the status of a governmental ministry, specifically responsible for environmental issues and natural resource protection. The SENMA will formulate national conservation policies, and coordinate and regulate the activities of other resource management institutions, to ensure compliance. All protected areas in the country are to be unified into a national system under the management system of SENMA. At the local level, Departmental Secretariats of the Environment will be responsible for natural resources and protected areas in each department. Department Councils of the Environment are to be established, to formulate local conservation policies and assess the activities of the Department Secretariats.

The project for General Environmental Law proposes to unify all existing protected areas in the country into the National System of Protected Areas, under the administration of the proposed Secretariat of the Environment. However, standard definitions of management categories to be included in the system are not given, but are to be stated in legislation providing for the creation of each protected area.

3. Extent, Status and Security of Production Forests

The Forest Law regulates the exploitation of forest resources and makes provisions for the creation of forest reserves for the protection and conservation of forests. The designations include: 1) Permanent production forest; 2) Permanent protection forest; 3) Closed forest reserve; 4) Special forest; and 5) Multiple-use forest. There are two types of unclassified forest: 1) National fiscal forest on unframed land; and 2) unclassified under private ownership.

According to the Bolivian Conservation Data Center (CDC), there are five production forests in Bolivia which total 6.4 million ha. Management authority for these forests resides with the Forest Department which is under the jurisdiction of the CDF. The 1986 *Environmental Profile of Bolivia*, published by the United States Agency for International Development in collaboration with the Government of Bolivia, described serious institutional weakness within the CDF. The report went on to criticise the CDF for its total lack of research to orient planning activities towards the sustainable management of forests, their orientation towards short-term gains and their policy of allowing settlement within the boundaries of production forests. The *Environmental Profile* describes granting of timber concessions in the forest reserves of Chore and Guarayos as having been indiscriminate and, in some cases, fraudulent.

A Catalogue of Environmental Legislation, recently published by CDC, contains a list of 830 provisions for environmental management. Among these, more than 300 refer to renewable natural resources (such as forest resources and wildlife) and protected areas.

The CDF has not coordinated their activities with other government agencies and this has led to conflicts. The National Colonisation Institute (INC), for instance, has settled people within forest reserves.

Forest exploitation in Bolivia is selective, based on the extraction of a few species such as mara (*Swietenia macrophylla*), cedro (*Cedrela* spp.), roble (*Amburana cearensis*) and ochoo (*Hura crepitans*). Typically, a large amount of usable residual material is left in the forest and secondary species are not exploited. One USAID-contracted study (CDF, CI) estimated that, on average, only 39% of exploitable timber volumes were removed in harvesting operations. The current timber fee structure, which charges on the basis of final production rather than stumpage, discourages the efficient use of forest resources. Timber exploitation is conducted without the benefit of silvicultural or ecological studies. Post-logging surveys are not prepared.

Because of the highly selective nature of timber operations in Bolivia, the ecological impact of logging is relatively low. The problem is that because of deficient and sometimes non-existent forest administration, supervision and control, logged areas become entry points for hunters and illegal settlers. In some areas, clandestine logging has become an important underground industry. The illegal loggers, or "cuartoneros" cut boards in the forest with chain saws. This method of harvesting causes great waste of the resource.

Forest concessions in Bolivia occupy 22.5 million ha. This represents almost 50% of total forest cover. Bolivian law requires all concessions to be located in production forests or, in some exceptional cases, in unclassified forests. Nonetheless, the production forest estate totals only 6.4 million ha and most of the concessions are located in unclassified forests.

All concessionaires are required to produce management plans which include a timber inventory and descriptions of harvesting and regeneration systems which will be employed. The majority of concessionaires have conducted inventories and several have produced working plans as well. In practice, however, management prescriptions have not been followed, particularly with regard to diameter limits, annual allowable cut, parameters and area restrictions. Most companies are

Bibliography

Brockman, C.E. 1986. *Perfil Ambiental de Bolivia.* USAID. La Paz.

Brockman, C. (Ed.) 1978. *Mapa de Cobertura y Uso Actual de la Tierra.* ERTS, escala 1:1,000,000.

Cardozo, A. 1988. *Areas Protegidas de Bolivia.*PRODENA, La Paz, Bolivia. 86pp.

CDC-Bolivia, HNB, MNHN, CIEC, and CI.1988. *Diagnóstico de la Diversidad Biológica de Bolivia.* AID/Bolivia, Vol.1:143pp.Vol.2:100pp.

CDC-Bolivia. 1990. Rol de la Actividad Forestal en la Conservación de la Biodiversidad Biologica. Unpublished report prepared for the IUCN Forest Conservation Programme workshop, "Realistic Strategies for Tropical Forest Conservation" in Perth, Australia.

CDF, PNUD, FAO, PNUMA. 1989. *Plan de Acción para el Desarrollo Forestal. 1990-1995.*99pp.

Ellenberg, H. 1981. Mapa simplificado de las ecoregiones de Bolivia. *Ecologia en Bolivia* 1. La Paz, Bolivia

Freeman, P.H. *et.al.* 1979. Perfil ambiental de Bolivia. Un reconocimiento de campo. Informe preliminar. JRB Associates, Inc., McLean, Virginia.

Grimwood, I.R. and Whitmore, T.C. 1978. Bolivia. IUCN, Gland, Switzerland. Unpublished report. 29pp.

Hanagarth, W.and Arce, J.P. 1986. La situación de los Parques Nacionales y Reservas de Vida Silvestre en el Departamento de La Paz, en el marco de una planificación regional. *Ecologia en Bolivia* 9:1-67. 3 Maps.

Hanagarth, W. and Marconi, M. 1986. Parques Nacionales y áreas equivalentes. In: Brockman C.(Ed.). *Perfil Ambiental de Bolivia.* USAID-IIED. pp.36-55.

Hanagarth, W. 1988. Plan de acción forestal de los trópicos:Informe del consultor en manejo de protegidas, fauna silvestre y recursos genéticos. Bolivia. Unpublished.

IUCN. 1990. *1990 United Nations List of National Parks and Protected Areas.* IUCN, Gland, Switzerland and Cambridge, UK.

Jungius, H. and Pujol, R. 1970. *Bolivian National Parks and Reserves.* UNESCO, Serial No.1944/BMS. RD/SCE.Paris. 120pp. Unpublished.

Marconi, M. 1989. Base legal del sistema de areas protegidas de Bolivia. CDC/IT.44pp. Unpublished report.

Marconi, M. 1988. Areas protegidas. In: CDC-Bolivia. *Diagnóstico de la Diversidad Biológica de Bolivia.* pp.78-91.

Mittermeier, R.A. 1988. Primate diversity and the tropical forest.In: Wilson, E.O. 9th edn. 1988. *Biodiversity.* National Academy Press. pp.145-154.

Ormazabal, C. 1988. *Sistemas Nacionales de Areas Protegidas en America Latina.* Proyecto FAO/PNUMA FP 6105-85-01. Doc. Tec. No. 13. 97pp.

Poore, D., Burgess, P., Palmer, J., Rietbergen, S. and Synott, T. 1989. *No Timber Without Trees: Sustainability in the Tropical Forest.* Earthscan, London.

Rios R, M.A. 1980. Parques nacionales y otras áreas protegidas en Bolivia. An unpublished report prepared for the IUCN Comision on National Parks and Protected Areas. Lima 1980. 80pp.

Schurholz, G. 1977. Estudio para el establecimiento de una reserva de selva tropical. Informe final sobre el proyecto IUCN/WWF No. 1309. Pilon Lajas, Bolivia. Unpublished report.

Solomon, J. 1989. Bolivia. In: Campbell, D.G. and Hammond, H.D. (Ed.) 1989. *Floristic Inventory of Tropical Countries*. N.Y.B.G.-WWF. pp455-463.

Suarez, M.O. 1986. *Parques Nacionales y Afines de Bolivia*. La Paz, Bolivia. 134pp.

Toledo, V. 1985. *A Critical Evaluation of the Floristic Knowledge in Latin America and the Caribbean*. Washington, DC. 78pp.

Unzueta, O. 1975. *Mapa Ecológico de Bolivia*. MACA, La Paz, Bolivia.

WCMC. 1992. *Protected Areas of the World: A review of national systems. Volume 4: Neartic and Neotropical*. IUCN, Gland, Switzerland and Cambrigde, UK. 268pp.

World Resources Institute. 1990. *World Resources: 1990-91*. Oxford University Press, New York.

BOLIVIA
TOTAL FOREST AREA

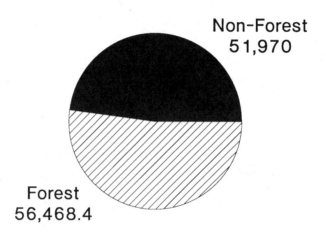

Non-Forest
51,970

Forest
56,468.4

Note: values given in 1000s of ha

BOLIVIA
LAND USE DESIGNATIONS

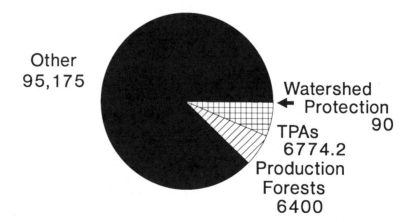

Other
95,175

Watershed
← Protection
90

TPAs
6774.2

Production
Forests
6400

Note: values given in 1000s of ha

BRAZIL

Based on the work of Herbert Schubart

1. Conclusions and Recommendations

The totally protected area network of Brazil covers a relatively small portion of national territory (2.4%). Coverage of this system is very uneven with most of the parks and protected areas located in those portions of the Amazon which have relatively low economic value. Critically threatened and highly diverse ecosystems in other parts of this country, particularly the Atlantic forests in the Northeastern, Southern and Southeastern Regions are under-represented.[1]

Although legislation covering environmental concerns is adequate, administrative capability to enforce existing laws is lacking. Population pressures and land use conflicts threaten the integrity of many TPAs.

Production Forest areas suffer from a similar lack of funding and manpower. Production forests (including both National Forests and Extractive Reserves) cover only 1.7% of Brazil and to date have not provided significant amounts of timber. Virtually all timber in Brazil comes from areas slated for conversion to alternative land uses.

Extractive Reserves are a newly recognised category of land use which have been created primarily for the benefit of rubber tappers. These areas have been sustainably managed for decades.

2. Present Extent, Status and Security of the TPAs

The totally protected area estate of Brazil contains 20,525,324ha or 2.4% of national territory. The TPA network includes several management categories. The principal divisions are National Parks, Biological Reserves, Ecological Stations and Environmental Protection Areas. Brazilian legislation describes these categories as follows:[2]

1. **National Parks**: Areas of exceptional natural attributes set aside with the purpose of combining integral protection of flora, fauna and natural beauty with utilisation for educational, recreational and scientific objectives. The principal objective is ecosystem preservation.

2. **Biological Reserves**: Strict reserves where activities which will modify the environment in any way are prohibited, with the exception of authorised scientific research.

---

1   The Brazilian Institute of Geography and Statistics (IBGE) defines five geographic regions as follows: **Northern Region** (Acre, Amapá, Amazonas, Para, Rondônia, Roraima and Tocantins); **Northeastern Region** (Alagoas, Bahia, Ceará, Maranhão, Paraiba, Pernambuco, Piauí, Rio Grande do Norte and Sergipe); **Southeastern Region** (Espírito Santo, Minas Gerais, Rio de Janeiro and São Paulo); **Southern Region** (Paraná, Rio Grande do Sul and Santa Catarina); **Central Western Region** (Mato Grosso do Sul, Districto Federal, Goiás and Mato Grosso).

2   Indian Reserves and Resource Reserves (which have a provisional status) are not considered here.

3. **Ecological Stations**: Areas representative of Brazilian ecosystems, destined for basic and applied ecological research, for protection of the natural environment and for development of conservation education.

4. **Environmental Protection Areas**: Corresponds to IUCN Category V, "Protected Landscapes and Seascapes". Nationally significant landscapes which are characteristic of the harmonious interaction of man and land which provide opportunities for recreation and tourism.

The 1988 Brazilian Constitution contains a chapter addressing environmental issues. The principles defined in this document include the conservation and restoration of ecological processes and the preservation of biological diversity. The constitution also mandates the creation of protected areas in all states.

The 1988 Constitution reinforces the Forest Code of 1965 which set criteria for the permanent preservation of forests, committed the government to create national parks and national forests and limited forest clearance on private lands in the Amazon to 50% of each plot.

The agency responsible for the control of parks and forests is the Brazilian Institute for the Environment and Renewable Natural Resources (IBAMA). IBAMA was created in 1989 by merging the two principal land management agencies, the Brazilian Forest Development Institute (IBDF) and the Special Secretariat for the Environment (SEMA). Two smaller agencies dealing with fisheries and natural rubber production were also included in the merger. This change resolves problems caused by lack of coordination and overlapping authority.

Even though Brazilian legislation provides ample protection for much of that country's natural heritage, problems exist. Administrative capability is generally lacking. The staff levels required for drafting and implementing management plans, demarcation of boundaries and controlling access are inadequate. Some of the older and smaller parks in the south and south-east receive modest funding and some level of protection. Many of the newer parks in the northern areas, however, have no park personnel assigned to them at all. In the legal Amazon, 73 park officers and guards are attempting to manage 13 million ha of TPAs.

Some of the TPAs in the Amazon are effectively protected by virtue of their isolation. Isolation, however, is not always a guarantee of protection as witnessed by the case of the Pico do Neblina National Park which has been invaded by gold miners. Wherever roads exist, TPAs come under pressure from illegal settlers.

Approximately 20-30% of the areas in the Amazon, which have been recently afforded legal protection, are actually privately owned. In order for these areas to be brought under state control, the government must purchase this land from private owners at an estimated cost of US$300 million.

The ecosystem coverage of the TPA estate is remarkably uneven. Most of the protected areas are located in the northern region which roughly corresponds to the Brazilian Amazon. Other regions of the country such as the *Araucaria* forests and the Atlantic forests are much more critically threatened. Native forests in the southeastern region, including secondary forests, have been reduced to only 4% of their original extent. Only 13% of the original forest cover in the southern region remains.

The non-Amazonian portion of Brazil has highly diverse ecosystems due to the occurrence of four broad morphoclimatic zones with complex transition zones between them. These areas are poorly protected.

Many of the legally protected areas were selected for opportunistic reasons or because of their low economic value. In 1982, the IBDF drafted the Plan for a Brazilian System of Conservation

Units, based on the IUCN classification system. This was designed to provide guidelines for TPA designation based on ecological criteria. Unfortunately this plan has not been implemented.

3. Extent, Status and Management of Production Forests

For the purposes of this study, the Production Forest Estate of Brazil is considered to include National Forests and Extractive Reserves. These legally recognised land use categories are defined as follows:

1. **National Forests**: Established by the Forest Code to provide a sustained production of timber and non-timber products.

2. **Extractive Reserves**: This conservation category was originally proposed by Brazil's Council of Rubber Tappers and the Rural Worker's Union to secure legal protection for forests traditionally used by them. These areas are used primarily for rubber tapping but other extractive activities are also allowed. The legal status of extractive reserves has not been completely settled.

There are four Extractive Reserves totalling 2,162,989ha. The thirty eight National Forests cover 12,598,852ha. This coverage represents an overall increase of 700% since 1988.

The National Forests suffer from the same problems as the protected area estate, that is to say that the management authority is underfunded and understaffed. National Forests are subject to pressures from landless farmers looking for new agricultural areas. The integrity of production forests is threatened and management levels are inadequate.

The Extractive Reserves are controlled by the rubber tappers themselves and are therefore better protected than National Forests. Violent land use conflicts have occurred between rubber tappers and people such as cattle ranchers who would like to see the reserves converted to alternative uses. Chico Mendes is the most famous victim of this violence but there have been many others. The government has had difficulty enforcing the law in these areas.

The rubber tappers are somewhat suspicious of the wisdom of timber management in natural forests. Their criticisms were summarised by Fearnside (1989) as follows:

1) The sustainability of natural forest management is still unproven.

2) The benefits to local people are often small.

3) Logging causes a high level of disturbance to the forest.

4) Top-down planning and administration makes management susceptible to changing official policies.

5) Management regimes are easily circumvented by corruption.

6) The reliance upon paid guards is a weakness in the face of migrants and speculators.

Nevertheless, because of a decline in the market value of natural rubber, proponents of the extractive reserve concept are exploring the possibility of including timber extraction in the management regimes of these areas (Perl *et al.*, 1991).

The supply of roundwood coming from National Forests is insignificant. Virtually all the timber produced in Brazil comes from land allocated for conversion to non-forest use except for plantations which supply wood used for charcoal, fuelwood, paper and cellulose. Currently, there is not a single example of a well established, commercial-scale project in Brazil.

There are some experimental forest management projects underway, however. First among these is the Tapajós National Forest in the state of Para. Eight years of research have yielded some positive results. Forest managers there are initiating a pilot project which aims to harvest 1000ha over a five year period. They will then apply lessons learned from this experiment to 132,000ha of forest which will be managed on a sustained yield basis.

In the Antimari State Forest the Technology Foundation of Acre (FUNTAC) is conducting an interesting experiment in multiple-use, sustained yield of a variety of products including rubber, Brazil nuts, timber, game, bamboo and medicinal plants.

In the National Forest of Caxiuanà in Belem, the Museo Paraense Emilio Goeldi is conducting research on sustainable forest management systems.

Basic Forest Statistics: Brazil

Total Land Area:

851,199,630ha	(IBGE, 1989. from: Schubart, 1990.)
845,651,000ha	(World Resources Institute, 1990. p.268)

Total Forest Area:

560,420,000ha	(WRI, 1990. p.268) ("Forest and Woodland" 1985-87)
514,480,000ha	(WRI, 1990. p.292) ("Extent of Forest and Woodland, 1980s: Total")

Deforestation Rate:

0.4%/yr	(derived from: Schubart, 1990. p.3)
1.8%/yr	(WRI, 1990. p.292)

Production Forest Estate:

12,598,852ha	(Schubart, 1990. p.13) (National Forests)
2,162,989ha	(Extractive Reserves)
14,761,841ha Total	(Units under federal administration only)
800,000	(Synnott, T. p.8) (Total productive and protective forests in tropical moist forest areas)

Totally Protected Area:

20,525,324ha	(IUCN, 1990. p.50)
16,377,719ha	(Schubart. 1990) (Units under federal administration only)
4,660,000ha	(WRI, 1990. p.292) ("Protected Closed Forest: 1980s")
20,096,133ha	(WRI, 1990. p.300) ("All Protected Areas")

Number of Units:

162	(IUCN, 1990. p.50)
92	(Schubart, 1990. p.11) (Units under federal administration only)
160	(WRI, 1990. p.300) ("All Protected Areas")

Bibliography

Fearnside, P.M. 1989. Extractive Reserves in Brazilian Amazona: an opportunity to maintain tropical rain forest under sustainable use. In: Schubart, H., *BioScience* 39(6):387-393.

IBGE (Brazilian Institute of Geography and Statistics) 1989. *Anuário Estadístico da Brazil.* Brazil, Rio de Janeiro, 49:1-716 from Schubert, H. 1990.

IUCN. 1990. *1990 United Nations List of National Parks and Protected Areas.* IUCN, Gland, Switzerland and Cambridge, UK.

Perl, M.A., Kiernan, M.J., McCaffrey, D., Buschbacher, R.J. and Batmanian, G.J. 1991. *Views from the Forest: Natural Forest Management Initiatives in Latin America.* Tropical Forestry Programme, World Wildlife Fund.

Poore, D., Burgess, P., Palmer, J., Rietbergen, S. and Synott, T. 1989. *No Timber Without Trees: Sustainability in the Tropical Forest.* Earthscan Publications Ltd, London.

Schubart, H. 1990. Commercial forestry and the conservation of biodiversity in tropical forests: the situation in Brazil. Unpublished report prepared for the IUCN Forest Conservation Programme workshop, "Realistic Strategies for Tropical Forest Conservation" in Perth, Australia.

Synnott, T. 1988. Natural Forest Management for Sustainable Timber Production: South America and the Caribbean. Unpublished report prepared for IIED and ITTO.

World Resources Institute. 1990. World Resources: 1990-91. Oxford University Press. New York.

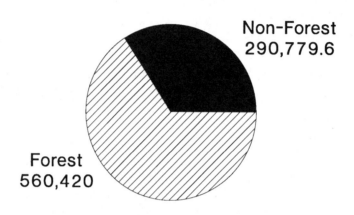

BRAZIL
TOTAL FOREST AREA

Non-Forest
290,779.6

Forest
560,420

Note: values given in 1000s of ha

BRAZIL
LAND USE DESIGNATIONS

Other
815,912.5

TPAs
20,525.3

Production
14,761.8

Note: values given in 1000s of ha

COLOMBIA

Based on the work of Julio Carrizosa Umaña

1. Conclusions and Recommendations

Colombia's totally protected area network covers approximately 9% of its national territory. This network contains only 44% of the ecosystem types found in Colombia. The integrity of these protected areas is severely threatened by the inability of the government to exercise effective state control in the countryside. Lack of funds, threats from guerrilla groups and the activities of drug traffickers conspire to make government-sponsored natural resource management extremely difficult.

Colombia has one of the highest levels of species diversity per unit area in the world and is probably second only to Brazil in overall diversity. Colombia's special status as a "megadiversity" country make its problems of conservation a special concern to the global community.

Colombia's production forest estate covers 1.5% of national territory. In some cases production forests are contiguous with totally protected areas and act as buffer zones. Some production forests are located in ecosystem types which are under-represented by the parks and reserves system.

Many changes in current land use trends will have to take place before production forests can contribute to biodiversity conservation. Current logging practices and post-harvest land use have led to severe ecological degradation and loss of biological diversity.

2. Extent, Status and Security of TPAs

Of Colombia's total land areas of 103,870,000ha, 9,016,893ha enjoy legal status as totally protected areas. Effective protection, however, is severely constrained by the lack of state control. Severe shortages of funds and manpower and the activities of guerrillas and drug traffickers make the exercise of sovereign control of Colombia's interior extremely problematic.

A few natural areas have been protected by private organisations such as the Fundación Pro-Sierra Nevada de Santa Marta that have been able to attract outside funding. The protection of some important water catchment areas has been financed by water authorities or from electricity fees in areas where they affect hydroelectric plants.

Coverage of various ecosystems remains incomplete. Only 44% of Colombia's ecosystem types are covered by the totally protected area system. The remainder are partially covered by the production forest areas.

3. Extent, Status and Security of Production Forests

In Colombia, 1,562,000ha of forest have been legally set aside for timber production. The laws governing the concession areas are generally coherent and adequate. In addition to fees for logging rights, concessionaires are required to pay fees for reforestation and security to protect the logging area from illegal incursions. They are required to report on a semester basis on the condition of the area under their control.

The problem is found less with the current laws than with the inability of government institutions to exercise effective control over national territory. Administrative capacity is so limited that the forestry department is unable to fully process the semester reports filed by the concessionaires, much less to adequately supervise timber operations.

Industrial forest operation are followed by small and medium scale loggers who are not subject to state control. These are, in turn, followed by local people who harvest the second-growth saplings and small diameter trees which find a ready market in the mining industry. This is a generalised practice which, logging concessionaires claim, makes it difficult to comply with government requirements to replace the exploited forests.

Experience in the management of natural forests is very limited in Colombia. In Carare, there is an ongoing experiment in enrichment planting. In the private sector, the company CARTON de Colombia has had some success in the management of natural forests for timber. A World Wildlife Fund review of the project reported that "from a purely technical point of view, the project demonstrates that it is possible to manage the forest to obtain a profit and to maintain satisfactory regeneration" (Perl *et al.*, 1991). Some reports, however, indicate that their silvicultural systems severely alter the structure and composition of the forest, thereby degrading its value as a reservoir of biological diversity. In addition, they have had trouble controlling illegal timber harvesting as many local people exploit trees within the concession area thus disrupting the production cycle prescribed in the management plan.

Basic Forest Statistics: Colombia

Total Land Area:
103,870,000ha   (World Resources Institute, 1990. p.268)

Total Forest Area:
43,000,000ha   (Carrizosa, 1990. p.1) (as of 1984)
38,700,000ha   (Carrizosa, 1990. p.2) (estimate based on deforestation rates since 1984)
53,100,000ha   (DNP, 1987)
51,570,000ha   (WRI, 1990. p.268) ("Forest and Woodland" 1985-87)
51,700,000ha   (WRI, 1990. p.292) ("Extent of Forest and Woodland, 1980's: Total")

Deforestation Rate:
1.0%/yr   (Carrizosa, 1990. p.2)
1.7%/yr   (WRI, 1990. p.292)

Production Forest Estate:
1,562,000ha   (Carrizosa, 1990. p.6)

Watershed Protection Forests:
333,000ha   (DNP, 1987)

Totally Protected Area:
9,016,893ha   (Carrizosa, 1990.)
9,301,690ha   (IUCN, 1990. p.75)
5,613,965ha   (WRI, 1990. p.300) ("All Protected Areas")
2,280,000ha   (WRI, 1990. p.292) ("Protected Closed Forests: 1980s))

Number of Units:
42   (Carrizosa, 1990.) (40 equal to or over 1000ha)
42   (IUCN, 1990. p.75)
35   (WRI, 1990. p.300)

Bibliography

Becerra, J.E. and Germán, M.G. 1972. Regeneración natural del roble después de la tala rasa y ensayo comparativo de diferentes sistemas de talar estas especies. Instituto de Investigaciones y Proyectos Madereros, Bogotá. Unpublished report.

Carrizosa Umaña, J. 1990. Papel de la Actividad Forestal en la Conservación de la Productividad Biologica: Colombia. Unpublished report prepared for the IUCN Forest Conservation Programme workshop, "Realistic Strategies for Forest Conservation" in Perth, Australia.

Carton Columbia. 1985. Reunión anual de investigación forestal. Cali, Columbia. Unpublished report.

Departamento Nacional de Planeación (DNP) 1987. *Plan de Acción Forestal*. Bogotá, Columbia.

Escobar, Milagro. 1981. Análisis estructural, estudio de la regeneración y tratamientos silviculturales en un bosque de catival. Empresas madereras del Atrato, Medellin. Unpublished report.

Gentry, A. Especies richness and floristic composition of Choco region plant communities. *Caldasia*. Octubre de 1986. Bogotá.

IGAC. 1984. *Bosques de Columbia*. Bogotá. Columbia.

IUCN. 1990. *1990 United Nations List of National Parks and Protected Areas*. IUCN, Gland, Switzerland and Cambridge, UK.

Labrach, W., Mazuera, H. 1985. Proveniencia y característica de la regeneración natural en un bosque en Cartón de Columbia. Cali, Columbia. Unpublished report.

Perl, M.A., Kiernan, M.J., McCaffrey, D., Buschbacher, R.J. and Batmanian, G.J. 1991. *Views from the Forest: Natural Forest Management Initiatives in Latin America*. Tropical Forestry Programme, World Wildlife Fund.

Poore, D., Burgess, P., Palmer, J., Rietbergen, S. and Synott, T. 1989. *No Timber Without Trees: Sustainability in the Tropical Forest*. Earthscan, London.

Salas, Gonzálo de las. 1981. Muestreo de la regeneración natural después del aprovechamiento dos áreas de tres y cinco años de edad en la concesión del Bajo Calima. CONIF, Bogotá.

Tosi, J. 1978. Observaciones sobre la ecologia de las áreas propuestas para la investigación sobre regeneración natural y artificial de bosques en la región Pacifica de Colombia. INDERENA FAO, PNUD.

World Resources Institute. 1990. *World Resources: 1990-91*. Oxford University Press, New York.

COLOMBIA
TOTAL FOREST AREA

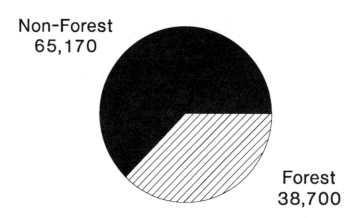

Non-Forest
65,170

Forest
38,700

Note: values given in 1000s of ha

COLOMBIA
LAND USE DESIGNATIONS

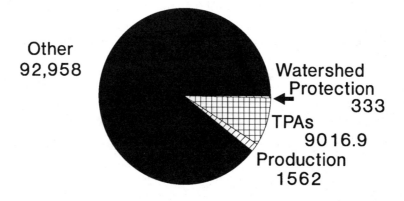

Other
92,958

Watershed
Protection
333

TPAs
9016.9

Production
1562

Note: values given in 1000s of ha

ECUADOR

Based on the work of Luis Suárez

1. Conclusions and Recommendations

Ecuador possesses a potentially valuable protected area estate covering 10% of the national territory and most ecosystem types. Important gaps in the coverage exist, however, particularly in the "Costa" region in the western part of the country. This deficiency may be remedied by proposals pending for additions to the parks and reserves system.

Although extensive, the TPA network is virtually unprotected. The parks administration is seriously underfunded and understaffed. Political support for protected areas is slight.

The Forest and Vegetation Protection Areas, which cover approximately 5% of Ecuador, could function as additional refugia for plant and animal species. Despite their legal protection, these areas face pressures from conversion to agriculture.

There is no Permanent Production Forest in Ecuador. The government abandoned their concessionaire system because of serious management and protection problems. Loggers are now given short-term licenses to extract timber from unreserved forests which have no special legal status and which are not protected or managed.

Some benefits might accrue to forest management activities if the timber fee structure was changed so as to eliminate incentives for wasteful logging practices. If the currently low fees were raised, increase revenues which could then be invested in protection and management activities.

The institutions dealing with production forests and parks must be significantly reorganised and strengthened. Legislation dealing with natural resource management needs to be redrafted so that it is consistent and effective. Enforcement structures have to be put in place.

On-going international initiatives such as the USAID SUBIR project and the recent debt-for-nature swap may improve Ecuador's capacity to protect and manage its natural resources.

2. Extent, Status and Security of TPAs

The terrestrial TPA estate in Ecuador encompasses 3,007,664ha. The national system of protected areas contains fifteen units including six National Parks, three Ecological Reserves, one Biological Reserve, two National Recreational Areas, two Fauna Production Reserves ("Reservas de Producción Faunistica") and one Geobotanical Reserve. The practical and legal distinctions between management units are not clearly defined.

The law allows for limited hunting activity in the Fauna Production Reserves. For this reason these two management units cannot technically be considered as totally protected areas as defined by IUCN categories I-V. The legally allowable hunting, however, is limited to the traditional wildlife harvests of the Siona-Secoya indigenous group. In Ecuador, these areas are legally and administratively part of the "National System of Protected Areas". For these reasons, the Fauna Production Reserves are included here as part of the TPA estate. Together these areas total 312,000ha.

It is important to note that the eastern portion of the Cuyabeno Fauna Production Reserve has been subject to extensive oil exploration and extraction activities. Many settlers have occupied land made accessible by the oil companies' roads.

The TPA estate is complemented by the Forest and Vegetation Protection Areas. This category includes both private and public lands such as property held by universities, research stations and private individuals. The parcels which make up the system are small, usually less than 5000ha and often less than 200ha. They occupy land on steep slopes or areas which are otherwise not suitable for agriculture. In total they comprise 1,350,000ha but at least 300,000ha of this is not forested.

Conservation interests have used the official designation of "Protection Forest" ("Bosque Protector") as a strategy to conserve relic forests and watersheds which would otherwise be susceptible to threats from agrarian reform. The legal protection afforded these areas has not been backed up by effective enforcement. As a result, Protection Forests have been subject to land conversion. Most of these forests lack management plans.

The conservation status of the rest of the TPA estate, unfortunately, is not much better. The administration of protected areas is the responsibility of the Department of Natural Areas and Wildlife Reserves (DANRS) which is part of the Sub-secretary of Forestry and Renewable Resources (SUFOREN). SUFOREN, in turn, is part of the Ministry of Agriculture and Livestock (MAG).

DANRS suffers from the typical problem of lack of funds and qualified personnel. As an example, in 1984 the total parks budget was US$250,000. Internally generated fees, such as those charged for the profitable expeditions to the Galapagos Islands, accounted for two thirds of this amount. In general, public and political support for protected areas is lacking.

In recent years, income from the Galapagos National Park has increased to approximately US$560,000 per year. The Galapagos receives additional funds from conservation foundations. Surplus revenue is used for other parks and reserves which together generate only US$40,000 per year.

The threats faced by parks and reserves are numerous. They include invasion by colonists, illegal timber harvesting, fuelwood and charcoal production, clandestine hunting and fishing, conversion of mangrove areas to shrimp ponds, mining and petroleum activity. Parks and reserves in Ecuador are almost entirely undefended. Legislation governing their operation is poorly conceived and contradictory.

The petroleum industry has opened up many new roads which have provided access for thousands of settlers. In the Cuyabeno Fauna Production Reserve and the Limoncocha Biological Reserve, petroleum activity within reserve boundaries has led to the uncontrolled conversion of thousands of hectares by farmers who follow newly created roads.

Most, but not all life zones are represented. Some serious gaps exist in the TPA system. Most notably, the highly threatened forest ecosystems in the Costa region are underrepresented. Priority conservation areas include coastal mangroves, dry southern coastal forests and lowland wet coastal forests. The 1989 Strategy for the National System of Protected Areas, developed by the government of Ecuador and the Fundación Natura, identifies actions for improvement and suggests 17 additional areas for inclusion in the TPA estate.

Several international initiatives are underway which aim to support the conservation of natural resources in Ecuador. First among these is a ten year, US$15 million USAID project to promote the sustainable use of biological resources. The SUBIR project, as it is called, will be implemented by a consortium of three NGOs: The Nature Conservancy, CARE and Wildlife

Conservation International. In 1989, Ecuador was also the beneficiary of a US$3.61 million debt-for-nature swap sponsored by The Nature Conservancy and the Fundación Natura, the principal conservation NGO in that country. In 1987 and 1989, the World Wildlife Fund sponsored debt swaps of US$1,000,000 and US$5,400,000 respectively on behalf of conservation interests in Ecuador. Interest from this money will be used for various conservation activities including parks protection.

The Ecuadorian Forest Law recognises customary land-use rights. It directs MAG to establish boundaries for areas claimed by indigenous people and to protect these territories from colonisation and commercial exploitation. Administrative mechanisms exist for collaboration with indigenous groups in conservation initiatives. A collaborative venture of this type was developed with the Awa community in northwestern Ecuador. In places where the rights of native people conflict with oil and timber interests, enforcement of customary land-use rights has been weak. In the Cuyabeno Faunal Production Reserve, for instance, the law recognises the hunting and fishing rights of the Siona-Socoya people. In practice, these people have not been protected from outsiders who poach in their traditional territory. A more serious threat to the Cuyabeno reserve is posed by colonists who enter new areas opened by oil activities and then successfully claim indigenous lands by obtaining legal titles from the land reform agency, IERAC. There is concern that this pattern may be repeated in the Yasuní National Park along a proposed oil company road in the Huaorani territory.

3. Extent, Status and Security of Production Forests

There are no permanent production forests in Ecuador. Timber supplies in Ecuador come largely from unreserved forest lands ("tierras baldias") which are converted to agriculture. Some timber is extracted from unreserved forests using a system of short term licenses. These forests are government-owned lands which have no particular legal designation.

The majority of Ecuador's timber comes from the Costa (western) region particularly the Esmeraldas Province. The Oriente (eastern) is increasingly being logged but infrastructure problems hamper timber extraction.

The former system of long-term concessions operating in legally defined concession areas with prescriptions for annual yields was almost completely abandoned by the government in 1981. Reportedly only one such concession is still operating. The government has cancelled all other contracts. The concessionaire system was abandoned because loggers did not abide by the terms of their contracts and because of practical and political difficulties associated with protecting production areas from occupation.

Logging practices are inefficient and unsustainable. Felling, skidding and landing techniques are primitive and cause great damage to the residual stand. Timber royalties are based on the timber volumes removed from forests. This creates a disincentive for efficient use of timber resources. In the Napo province, for instance, 17 m^3/ha were removed versus a total of 130 m^3/ha of standing volume available. In addition to being poorly structured, timber fees are also extremely low, in the range of US$0.40 to US$1.0/m^3. The low fee precludes the possibility of significant revenues which might then be channelled back into management activities.

The Department of Forest Management of SUFOREN has little financial or political support. There is a severe shortage of qualified personnel. The existing staff composition is skewed towards university level graduates. Field technicians and mid-level administrators are quite scarce.

SUFOREN has no direct role in policy development and is only marginally influential in national level decision making spheres. It lacks a meaningful planning and policy and strategy making process and there are no evaluation and control mechanisms. Many agencies and institutions are involved in management and implementation of activities in forested areas, but there is an absence of a coordination mechanism.

There is a Forestry Action Plan for Ecuador. The plan acknowledges many of the problems described above and outlines the following objectives:

1)   Increase the participation of the forestry sector in the national economy.

2)   Increase the use of agroforestry systems.

3)   Improve watershed management.

4)   Promote the rational use and conservation of forest resources.

5)   Increase the benefits of forest utilisation to native people and local farmers.

6)   Promote higher awareness of conservation issues.

Even though the Forestry Action Plan for Ecuador proposes numerous conservation projects for tropical forests, it has been criticised by some conservation groups for its emphasis on exploitative activities.

Basic Forest Statistics: Ecuador

Total Land Area:

27,684,000ha   (World Resources Institute, 1990. p.268)

Total Forest Area:

11,473,000ha	(Suarez, 1990. p.1)
12,000,000ha	(Poore, 1989. p.84)
14,730,000ha	(WRI, 1990. p.292)
12,098,000ha	(WRI, 1990. p.268)
14,140,000ha	(IUCN, 1988)
11,500,000ha	(Min. of Agriculture, 1991)

Deforestation Rate:

2.3%	(Suarez 1990.)
Bet. 0.8% & 2.6%	(derived from: Suarez, 1990. p.1)
2.3%	(WRI, 1990. p.292)
2.4%	(IUCN, 1988.)

Production Forest Estate:

3,000,000ha   (Synnott, 1988. p.8) (Tropical moist forest areas only. Includes both production and watershed protection forests)

Watershed Protection Forests:

1,350,000ha   (Suarez, 1990. p.3) (These are "Forest and Vegetation Protection Areas" created primarily to protect watersheds. At least 300,000ha of this, however, is not forested.)

Totally Protected Area:

3,007,664ha	(Derived from: IUCN, 1990. and including data on Chimborazo and Cuyabeno Fauna Production Reserves from: Cifuentes, 1989.)
2,865,477ha	(Suarez, 1990. p.2)
10,685,664ha	(IUCN, 1990. p.83)
2,695,664ha	(IUCN, 1990. p.83) (not including the 7,990,000ha Galapagos Marine Reserve)
10,685,664ha	(WRI, 1990. p.300) ("All Protected Areas")
350,000ha	(WRI, 1990. p.292) ("Protected Closed Forest: 1980s")

Number of Units:

15	(Suraez 1990)
14	(IUCN, 1990. p.83) (13 w/o Galapagos Marine Reserve)
13	(WRI, 1990. p.300)

Bibliography

Cifuentes, M. *et al.* 1989. *Estrategia para el Sistema Nacional de Areas Protegidas del Ecuador, II Fase*. IUCN. Quito, Ecuador.

IUCN. 1988. Ecuador: Conservation of Biological Diversity. Conservation Monitoring Centre, Cambridge, UK.

IUCN. 1990. *1990 United Nations List of National Parks and Protected Areas*. IUCN, Gland, Switzerland and Cambridge, UK.

Ministerio de Agricultura y Ganaderia. 1987. Situación forestal del Ecuador: resumen ejecutivo. Ministerio de Agricultura y Ganaderia, Dirección Nacional Forestal, Quito, Ecuador.

Ministerio de Agricultura y Ganaderia. 1991. Forestry action plan: executive summary. Quito, Ecuador.

Poore, D., Burgess, P., Palmer, J., Rietbergen, S. and Synott, T. 1989. *No Timber Without Trees: Sustainability in the Tropical Forest*. Earthscan, London, UK.

Synnott, T. 1988. Natural Forest Management for Sustainable Timber Production: Latin America and the Caribbean. Unpublished report prepared for IIED and ITTO.

Suarez, L. 1990. El papel de la actividad forestal en la conservación de la diversidad biológica del Ecuador. Unpublished report prepared for the IUCN Forest Conservation Programme workshop, "Realistic Strategies for Tropical Forest Conservation" in Perth, Australia.

World Resources Institute. 1990. *World Resources: 1990-91*. Oxford University Press, New York.

ECUADOR
TOTAL FOREST AREA

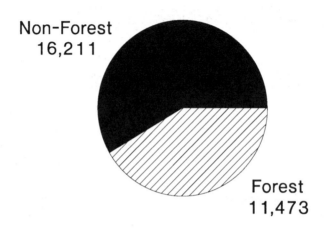

Non-Forest
16,211

Forest
11,473

Note: values given in 1000s of ha

ECUADOR
LAND USE DESIGNATIONS

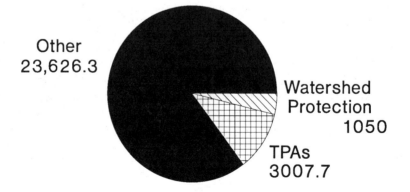

Other
23,626.3

Watershed
Protection
1050

TPAs
3007.7

Note: values given in 1000s of ha

HONDURAS

Based on the work of
José G. Flores Rodas and Edwin Mateo Molina

1. Conclusions and Recommendations

Recent innovations in the Honduran forestry sector may provide lessons for other countries. Reforms of the timber revenue system have led to improvements in tree utilisation and have allowed the forest service to become self-financing. This has taken place without undue harm to the timber industry. Problems still exist, however.

Over six percent of Honduras is legally protected. If the Honduran government approves pending legislation which mandates the creation of additional parks and reserves, 9.5% of Honduras will be set aside. The Honduran public is increasingly well informed about ecological issues and supportive of conservation initiatives. Honduran legislation is relatively enlightened with regard to protected areas. Recent reorganisation of the two main government institutions responsible for natural resource management are likely to improve administrative capabilities in this area.

The problems faced by Honduras are similar to those faced by other countries in the region. Funding restrictions and the lack of well trained personnel hamper the effectiveness of parks and protected areas management. Population pressures, uncontrolled conversion of forested areas and illegal logging seriously threaten the integrity of TPAs.

In the forestry sector, natural resource managers are facing the same problems. The situation is improving as the result of the USAID funded Forestry Development Project which aims to strengthen the forestry institutions and improve forestry practices in the pine areas.

2. Extent, Status and Security of TPAs

There are currently 45 protected areas in Honduras. The Honduran Congress is considering adding 26 additional areas.[1] Of the 34 administrative units for which data is available, TPAs cover a total of 703,340ha. Many of the remaining 11 units are cultural monuments or relatively small areas. The 26 proposed areas represent over 351,000ha. Data on the extent of some of the proposed areas is not yet available.

This TPA estate represents a significant portion of the remaining forest cover which totals 5,051,000ha. Protection of these areas, however, is ineffective. Many lack management plans and have not even been delineated yet. Only a small number of TPAs have permanent, full-time personnel assigned to them. The integrity of most protected areas is being violated by ranching and agriculture. Lesser problems include hunting, gold panning and illegal trade in wildlife.

On the positive side of the conservation ledger, a very significant conservation movement is evolving in Honduras. Honduras was one of the first countries to create an agency responsible for environmental protection. They were also one of the first countries to adopt an ecological

---

1   Two additional proposed areas are Indian reserves.

approach to the planning of their forested areas. The Honduran National Development Plan has a strong environmental orientation and the conservation NGOs are increasingly professional and effective. Much of the actual management of protected areas is being conducted by private organisations. In 1990, the National Commission on Environment and Development (CONAMA) was formed to advise the executive branch on environmental issues.

Many parks staff have received training at the Center for Tropical Research and Education (CATIE) or the US Fish and Wildlife Service. Nevertheless, there is still a serious shortage of trained people needed to manage protected areas.

Until recently, there was no special legislation for protected area management. An institutional gap existed between the two main natural resource agencies, the Renewable Natural Resources General Directorate (RENARE) and the Honduran Forestry Development Corporation (COHDEFOR). Administrative responsibility of TPAs was unclear. This problem has hopefully been addressed by the recent absorption of RENARE by COHDEFOR.

Many parks and protected areas face serious protection problems including illegal logging and slash-and-burn agriculture. In some cases refugees from the Nicaraguan war settled within park boundaries. Although most of these people have returned to Nicaragua, the rehabilitation of former settlement areas poses a significant challenge.

Honduras is receiving support for the management of parks and protected areas from US Agency for International Development, US Peace Corps, Canadian International Development Agency, United Nations Development Programme, the government of Holland, World Wildlife Fund and IUCN – The World Conservation Union.

3. Extent, Status and Security of Production Forests

There are four "forest reserves" in Honduras covering a total area of 1,642,011ha. The largest of these, the Olancho Reserve, covers approximately 1.5 million ha and is divided into nine separate management units. The 480,000ha of pine forests within Olancho provide the country with 50% of its pine products and supply 13 sawmills.

COHDEFOR is the government agency responsible for all forest management. It has the authority to regulate all forests regardless of tenure. The Honduran government established COHDEFOR in 1974 as an autonomous state agency, one of the few in Latin America.

Although COHDEFOR has legal jurisdiction over all forests in Honduras, it has not been willing or able to exercise control over the expansion of the agricultural frontier or over most logging activity. Reportedly, extraction rates in the reserves exceed replacement rates and agricultural conversion is progressing rapidly.[1] COHDEFOR's presence at the local rural community level has not been effective. These problems are due largely to the lack of institutional capability and the lack of forestry personnel.

The most effective forest conservation effort, in terms of protection and management, is the Integrated Management Area (AMI) project which was initiated experimentally in 1983 in cooperation with FAO and funded by the government of Norway. The AMI was officially adopted as the smallest forestry administrative unit in 1986. It consists of areas between 5,000

---

1   Guevara Moncada *in litt.*, 1991

to 10,000ha and involves a minimum of 100 families. The initiative operates in areas with abundant forests and agriculture and where there have been difficulties managing the resource (i.e. wildfires, pests, grazing animals, illegal timber cutting etc.). A forester who is well trained in communication and extension is assigned to a AMI and begins gaining the trust of the families by introducing improved agricultural techniques. At a later stage, the extension worker will start promoting forest management activities which generate income. The goal is to make local communities guardians of the forest. The AMI scheme has been successful in slowing or eliminating deforestation.[1]

In another notable initiative, COHDEFOR management of the pine areas has been the focus of the US Agency for International Development funded Forestry Development Project. The project has three objectives:

1) Improving administrative capability, reforming policies and improving timber sale methods based on stumpage fees.

2) Providing technical and financial assistance to sawmills.

3) Managing of selected forest areas as pilot projects. These projects will include the drafting of management plans, protection measures, reforestation, watershed conservation and environmental impact assessments.

Forest managers in pilot project areas are implementing the following measures designed to mitigate negative environmental impacts:

1) Delineation of buffer strips near streams and rivers.

2) Implementation of environmental standards for road construction.

3) Careful design of skid trails with consideration for topography and slope.

4) Habitat protection, including the preservation of partially decayed trees used as habitat for fauna.

5) Erosion control measures.

6) The closing of roads in areas susceptible to colonisation.

Natural regeneration silvicultural systems are used wherever possible. Post-logging recovery is monitored by forest technicians.

One of the early successes of the Forestry Development Project has been the reform of the timber revenue structures. All pine timber sales in Honduras are now done on a stumpage basis ("standing timber sale"). Stumpage fees have trebled from 12.00 lempiras per cubic meter (US$2.26) in 1990 to 36 lempiras per cubic meter (US$6.80) in 1991. Additional fees are charged for sawn timber exports.

Reforming the timber revenue structure by charging on a stumpage basis and by raising fees has produced two desirable effects. First, tree utilisation has increased 25%. This has happened because loggers and saw mill owners take greater pains not to waste wood. Second, COHDEFOR is now operating in the black after many years of running deficits. These reforms have taken place without undue disruption of the timber industry. One negative effect has been an increase in timber poaching.

---

1   Guevara Moncada *in litt.*, 1991

The success of the reform of the Honduran timber revenue system is encouraging. The current changes being imposed by COHDEFOR make that institution more likely to be able to achieve both timber production goals and conservation objectives. Hopefully, some of the lessons learned by COHDEFOR can be translated to forest services in other parts of the world as well.

Basic Forest Statistics: Honduras

Total Land Area:

11,189,000ha	(World Resources Institute, 1990. p.268)
11,208,800ha	(Poore *et al.*, 1989. p.86)

Total Forest Area:

5,051,000ha	(Flores Rodas and Mateo Molina, 1990. p.1)
3,580,000ha	(WRI, 1990. p.268)
3,997,000ha	(WRI, 1990. p.292)
5,000,000ha	(SECPLAN, 1989a. p.21)

Deforestation Rate:

1.6%	(derived from: Flores Rodas and Mateo Molina, 1990. p.1)
2.3%	(WRI, 1990. p.292)
0.7%	(derived from: SECPLAN, 1989a. p.21)

Production Forest Estate:

1,642,011ha	(SECPLAN, 1989b. p.203)
58,000ha	(WRI, 1990. p.292)
250,000ha	(Poore *et al.*, 1989. p.79) (tropical moist forest reserves including production and watershed protection areas)
1,437,000ha	(Poore *et al.*, 1989. p.86) (Olancho Reserve only)

Watershed Protection Forests:

3,280ha	(SECPLAN, 1989b. p.203)

Totally Protected Area:

703,340ha	(derived from: IUCN, 1990. p.100 and SECPLAN, 1989b pp.200-203)[1]
580,369ha	(WRI, 1990. p.300)
709,369ha	(IUCN, 1990. p.100)
904,700ha	(Flores Rodas and Mateo Molina, 1990. p.2) (including proposed areas)

---

1   This figure is based on data from the *Perfil Ambiental de Honduras* pp.200-203 and excludes proposed areas, forest reserves (IUCN category VIII), watershed protection areas and the Rio Platano Biosphere Reserve (IUCN category IX). It includes the Rio Platano Park which is contained within the Biosphere Reserve. The figure also includes data for Santa Bárbara, Azul Meambar, and Islas de Bahia. The data on these three units is contained in the *UN List of National Parks and Protected Areas* but not in the *Perfil.*

Number of Units:

45	(Derived from: SECPLAN, 1989b. pp.200-203)
34	(Derived from: SECPLAN, 1989b. pp.200-203) (Includes only units for which data is available)
15	(WRI, 1990. p.300)
34	(IUCN, 1990. p.100)
51	(Flores Rodas and Mateo Molina, 1990. p.2) (existing)
28	(Flores Rodas and Mateo Molina, 1990. p.2) (proposed)

Bibliography

Flores Rodas, J.G. and Mateo Molina, E. 1990. Study of the role of biodiversity conservation in Honduras. Unpublished report prepared for the IUCN Forest Conservation Programme workshop, "Realistic Strategies for Tropical Forest Conservation" in Perth, Australia.

Howard Borjas, P. 1988. *Impacto de la Expansión Ganaderia en la Crisis Alimentaria y el Desempeleo y Subempleo Rural y Alternativas de Emergencia.* SECPLAN/OIT/FNUAP/ HON/Po2, Tegucigalpa, Honduras.

IUCN. 1990. *1990 United Nations List of National Parks and Protected Areas.* IUCN, Gland, Switzerland and Cambridge, UK.

Poore, D., Burgess, P., Palmer, J., Rietbergen, S. and Synott, T. 1989. *No Timber Without Trees: Sustainability in the Tropical Forest.* Earthscan Publications Ltd, London.

SECPLAN. 1989a. *Perfil Ambiental de Honduras 1989: English Summary.* USAID. Washington, DC.

SECPLAN. 1989b. *Perfil Ambiental de Honduras 1989.* USAID. Washington, DC.

World Resources Institute (WRI). 1990. *World Resources: 1990-91.* Oxford University Press. New York.

HONDURAS
TOTAL FOREST AREA

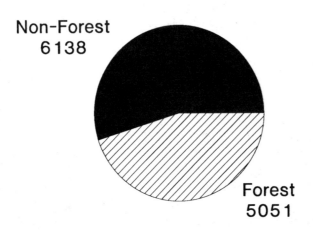

Non-Forest
6 138

Forest
5051

Note: values given in 1000s of ha

HONDURAS
LAND USE DESIGNATIONS

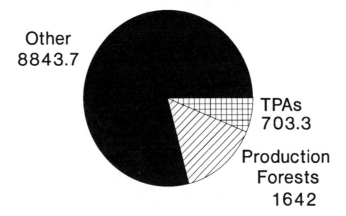

Other
8843.7

TPAs
703.3

Production
Forests
1642

Note: values given in 1000s of ha

PANAMA

Based on the work of Eric Rodriguez

1. Conclusions and Recommendations

Approximately 17% of Panama's total area of 7,708,000ha has total legal protection. Unfortunately, because of lack of institutional capacity, funding shortages, and population pressures, protection laws have not been fully implemented.

Panama's production forest estate, which covers 2.8% of the country, has also suffered from inadequate management and has not been protected. Forty-three percent of the legally recognised production forest area has been converted to agriculture and cattle pasture. Much of the remainder is degraded.

The Forestry Action Plan of Panama (PAFP) addresses many of the conservation problems faced by that country. Panama's conservation initiatives have received significant levels of support from many international organisations.

2. Extent, Status and Security of TPAs

The current totally protected area estate in Panama covers 1,332,140ha. This represents 17% of Panama's total land area. Although the government of Panama has been willing to designate protected areas, it has been unwilling or unable to commit funds for demarcation and management. As a result, many parks exist only on paper. Several protected areas have management plans but few of these are effectively implemented. Legislation governing protected areas is out of date, incomprehensible and inadequate according to the 1990 Forest Action Plan for Panama (PAFP).

The discouraging condition of the conservation situation is improving as Panama becomes the focus of several internationally supported initiatives to improve the management of protected areas. First among these is an ongoing USAID natural resource and institution-building project which will eventually total US$30-40 million. International multilateral and non-governmental organisations such as the Food and Agriculture Organisation (FAO), Unesco, the Interamerican Institute for Agricultural Cooperation (IICA), The Centre for Tropical Agricultural Research and Investigation (CATIE), the Japanese International Cooperative Agency (JICA), IUCN – The World Conservation Union, World Wide Fund for Nature (WWF) and The Nature Conservancy are also providing significant levels of technical support and financing for conservation activities.

International interest in the wise stewardship of Panama's natural resources is based on several factors. Panama occupies a land bridge between North and South America. As such it contains within its borders unique ecological communities with characteristics of both southern and northern hemispheres. In addition, Panama is located at the convergence of three of the four main bird migration routes in the Americas.

Panamanians have a vested economic interest in the preservation of tree cover in the watershed which drains into the canal zone. The erosion and siltation which results from the conversion of these areas to other uses directly threatens the long-term viability of the canal system which is their primary income generator.

Hydroelectric resources such as the Bayano Hydroelectric Plant are also threatened by unplanned land conversion and soil erosion.

The Forest Action Plan of Panama (1990) calls for increased enforcement of the integrity of parks and protected areas. The plan outlines measures for the institutional strengthening of the National Institute for Renewable Natural Resources (INRENARE), the principal land management agency.

3. Extent, Status and Security of Production Forests

Of Panama's 3.3 million ha of remaining forests, 217,309ha have officially been set aside as production forests. The exploitation of these areas has been proceeding in an unplanned and haphazard fashion. Logged areas are not protected and therefore become entry points for hunters and illegal settlers. Fourty-four percent of the legally recognised production forest estate has been converted for agriculture or cattle ranching. Much of the remainder is highly degraded. In some cases, only the older residents remember that these areas were ever forest reserves.

Legislation governing the management and disposition of commercial forests was drafted by the government in 1966. This legislation provided for the creation of a national forest service and broadly defined its institutional mandate. Unfortunately, the specific regulations required to enforce this mandate were never drafted. As a result there has been significant legal and bureaucratic confusion over the administrative mechanisms for natural resource management in Panama. The Forestry Action Plan for Panama calls for improved forestry legislation.

In the production forestry sector, the PAFP aims to bring 400,000ha of natural forest under a regime of sustainable management for timber and non-timber resources. The Plan also calls for the reforestation of 62,000ha which will include both industrial plantations, small woodlots and agroforestry systems.

Many production forest areas are contiguous with totally protected areas and could form buffer zones if they were managed adequately. Currently, these areas are being degraded and converted.

Basic Forest Statistics: Panama

Total Land Area:

7,708,000ha	(Rodríquez, 1990. table 2)
7,599,000ha	(World Resources Institute, 1990. p.268)
7,700,000ha	(INRENARE, 1990b. p.2)

Total Forest Area:

3,300,000ha	(INRENARE 1990b. p.3)
2,764,443ha	(Rodríquez, 1990. table 2)
3,990,000ha	(WRI, 1990. p.268) ("Forest and Woodland" 1985-87)
4,165,000ha	(WRI, 1990. p.292) ("Extent of Forest and Woodland 1980s: Total")
3,182,166ha	(INRENARE, 1990a. p.38)

Deforestation Rate:

1.0%	(derived from: Rodríquez, 1990. p.1)
0.9%	(WRI, 1990. p.292)
2.1%	(derived from: INRENARE, 1990b.)

Production Forest Estate:

217,309ha	(Rodriguez, 1990. table 6) (Forest Reserves) (Only 123,000ha of these legally recognised production forests remain forested)
1,300,000ha	(INRENARE, 1990b. p.3)
4,883,289ha	(INRENARE, 1990a. p.38) (This includes all lands legally recognised as suitable for timber exploitation. Only 2,734,116ha, however, are forested)

Watershed Protection Forests:

351,000ha	(Rodriguez, 1990. table 6)
2,000,000ha	(INRENARE, 1990a. p.3) (this encompasses the TPA estate)

Totally Protected Area:

1,332,140ha	(Derived from: IUCN, 1990. and Rodríquez, 1990)
1,311,382ha	(WRI, 1990. p.300) ("All Protected Areas")
1,326,140ha	(IUCN, 1990. p.144)
1,056,259ha	(Rodríquez, 1990. table 6) (Nat.Parks, Wildlife Refuges and National Monuments)
1,386,000ha	(derived from: INRENARE, 1990b. p.3)
666,902ha	(WWF, 1988.)

Number of Units:

15	(derived from: IUCN, 1990 and Rodríquez, 1990[1])
14	(WRI, 1990. p.300)
16	(IUCN, 1990. p.144)
14	(Rodríquez, 1990. table 6)

---

1 This figure is derived from the *1990 UN List of National Parks and Protected Areas* and information from the study done by Rodriquez (tables 3 and 4). It only includes terrestrial parks greater than 1000ha. The figure also includes data on the El Cope National Park which is contained in the Rodriquez report but not the *UN List*.

Bibliography

Aceres, G., Durran, E., and Tovar, D. Enfoque sobre las acciones esirategicas de la fundación de parques nacionales y medio ambiente en apoyo a la gestión ambiental en Panama. Fundación Panama, Panama, 1986. Unpublished report.

Instituto Geografico Nacional Tommy Guardia. 1988. *Atlas de la Republica de Panama*. IGNTG, Panama.

Instituto National de Recursos Naturales Renovables (INRENARE). 1980. *Informe Nacional Forestal (1985-1989)*. INRENARE, Panama.

INRENARE. 1988. *Propuesta de Plan de Accion Forestal 1989-1993*. PNUD/FAO PAN-87-001. Panama.

INRENARE. 1990a. *Plan de Accion Forestal de Panama. Documento Principal*.

INRENARE. 1990b. *Forestry Action Plan of Panama*. Executive Summary.

INRENARE. 1990c. *Agenda Ecologica 1990-1994*. INRENARE, Panama.

IUCN. 1990. *1990 United Nations List of National Parks and Protected Areas*. IUCN, Gland, Switzerland and Cambridge, UK.

Poore, D., Burgess, P., Palmer, J., Rietbergen, S. and Synott, T. 1989. *No Timber Without Trees: Sustainability in the Tropical Forest*. Earthscan Publications Ltd, London.

Rodriquez, E. 1990. Untitled report prepared for the IUCN Forest Conservation Programme workshop, "Realistic Strategies for Tropical Forest Conservation" in Perth, Australia.

World Resources Institute (WRI). 1990. *World Resources: 1990-91*. Oxford University Press, New York.

World Wide Fund for Nature. 1988. WWF Country Conservation Profiles: Central America. Unpublished report.

PANAMA
TOTAL FOREST AREA

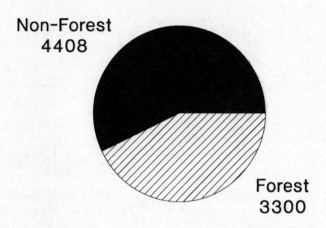

Non-Forest
4408

Forest
3300

Note: values given in 1000s of ha

PANAMA
LAND USE DESIGNATIONS

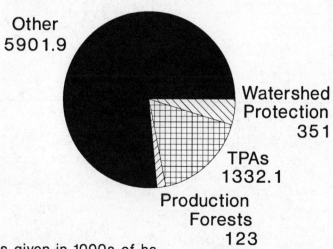

Other
5901.9

Watershed
Protection
351

TPAs
1332.1

Production
Forests
123

Note: values given in 1000s of ha

PERU

Based on the work of Gustavo Suarez de Freitas

1. Conclusions and Recommendations

The TPA network in Peru is relatively small and does not cover all ecosystem types. Prospects for biological diversity conservation could conceivably be enhanced by the large portion of forested estate which has been allotted to timber production. Although Peru has ample legislation regarding the protection and wise use of forests, administrative capabilities are not adequate to ensure that laws are enforced. Terrorist activity and well-financed drug traffickers have made conservation work difficult in certain regions. Logging, agriculture and pasture conversion take place without much regulation by the government and often with little regard for environmental safeguards and land capabilities.

Management of the production forests could be improved by revamping the system of timber fees to more effectively capture the portion of rent which should accrue to the government. These increased funds could then be channelled back into management and protection activities. Government incentives which encourage deforestation, such as land clearing requirements for acquiring title to land, should be eliminated. Financial incentives for management and reforestation should be created. The number of forest concessions should be reduced to a level that could be effectively managed by government authorities.

2. Extent, Status and Security of TPAs

The total land area of Peru is 128,521,600ha of which 75,686,000ha, or 59%, is forested. Most of the forest estate is located in the Peruvian Amazon. According to the National Forest Action Plan (1987) 73,700,000ha of forest are located in the Amazon Basin.

In Peru 5,517,835ha of land has been legally protected. This only includes totally protected areas such as Parks and National Sanctuaries (IUCN categories I-V).

Many of Peru's TPAs are effectively protected by their isolation from population centres and roads. The principal threats to TPAs come from spontaneous agricultural settlement in more accessible areas. This threat has abated somewhat due to the inability of the Peruvian government to finance road and colonisation projects. The degradation or destruction of TPAs and other forested areas will increase as more roads are built. The proposed road in the south of Peru connecting the country with Brazil, which was to have been financed by the Japanese government, posed a particular threat.

The coastal and mountain ecosystems in Peru are severely degraded and under-represented in the TPA network.

3. Extent, Status and Security of Production Forests

The government of Peru has set aside 42,214,000ha of forests for the permanent production of timber. This represents 60% of the forested areas in Peru and 33% of total land area. The relatively large area devoted to exploitative activities contrasts with the smaller portion of national territory which has been afforded protected area status (4%).

The production segment of Peru's forest estate is divided into two categories, National Forests (5,514,000ha) and Free Availability Forests (36,700,000ha) ("Bosques de Libre Disponibilidad"). In National Forests concessionaires receive 20 year renewable contracts on parcels which range from 20,000 to 200,000ha. In order to be awarded a concession, applicants must provide a technical/economic feasibility study. In practice, this feasibility study is a bureaucratic formality rather than an effective instrument for sound forest management.

Concessions in Free Availability Forests are available up to 100,000ha. In reality, the majority of contracts are for areas less than 1000ha because concessionaires can be awarded parcels of this size without having to file feasibility studies.

Peruvian forestry legislation is quite comprehensive but it has not been effective due to lack of enforcement capability. For instance, the Forestry and Wildlife Law requires logging enterprises to reforest. Loggers are reluctant to abide by their legal requirements because they lack the technical ability to undertake silvicultural operations, the costs of reforestation are high, and reforested areas can not be protected from colonists. The low fees charged for timber do not cover the costs of forest management.

Loggers engage in forest exploitation without technical supervision and outside the control of any effective national institution. The very large number of small contractors in the Free Availability Forests (3300 in 1986) make effective supervision virtually impossible.

The fees charged by government authorities are very low and do not even cover minimum costs of management and post-logging treatments. Fees are structured in such a way as to discourage optimal use of timber resources.

The permanence of the forest estate is seriously jeopardised by uncontrolled, spontaneous colonisation of forest areas. This problem is exacerbated by government requirements for "land improvement" (i.e. deforestation) for land titles and credit.

Some promising experiments in natural forest management are underway in Peru. For instance, the Palcazu Project located in the Central Forest is a notable example of sustainable forest management. The project operates in an area controlled by people from the Yanesha ethnic group and uses a strip harvesting system. In the Von Humbolt Forest, there has been an attempt to implement an integrated forest management system. Forest managers there are experimenting with natural regeneration techniques and have been able to achieve gains in harvesting efficiency.

These and other experiences in natural forest management demonstrate the feasibility of improved silvicultural techniques. Lessons learned from these projects could be transferred to other areas in Peru.

The very large proportion of Peruvian territory currently dedicated to production forestry could provide an opportunity to further the objective of biological diversity conservation if it were managed properly. Currently, however, logged areas are significantly degraded and are not effectively protected from clearance for agriculture.

Basic Forest Statistics: Peru

Total Land Area:

128,521,000 ha	(Suarez de Freitas, 1990. p.1)
128,000,000 ha	(World Resources Institute, 1990. p.268)
128,512,600 ha	(Poore *et al.*, 1989. p.83)

Total Forest Area:

75,686,000ha	(Suarez de Freitas, 1990. p.1)
69,400,000ha	(WRI, 1990. p.268) ("Forest and Woodland" 1985-87)
70,640,000ha	(WRI, 1990. p.292) ("Extent of Forest and Woodland, 1980s: Total")
69,680,000ha	(Poore *et al.*, 1990. p.83) (tropical moist forest)
77,600,000ha	(FAO, 1981. from: Poore. 1989. p.83)

Deforestation Rate:

0.4%	(derived from: Suarez de Freitas, 1990. p.1)
0.4%	(WRI, 1990. p.292)

Production Forest Estate:

42,214,000ha	(Poore *et al.*, 1989. p.83)
40,123,688ha	(Suarez de Freitas, 1990. p.11) (Amazon only) (42 units)

Watershed Protection Forests:

387,818ha	(Suarez, 1990. annex 2) (Amazon only)
4,342,000ha	(Poore *et al.*, 1989. p.83)

Totally Protected Area:

5,517,835ha	(IUCN, 1990. p.146)
580,000ha	(WRI, 1990. p.292) ("Protected Closed Forest: 1980s")
5,482,935ha	(WRI, 1990. p.300) ("All Protected Areas")

Number of Units:

22	(WRI, 1990. p.300)
24	(IUCN, 1990. p.146)

Bibliography

DGFF (1990). *Compendio de Estadísticas de Forestal y Fauna 1980-1989.*

Dourojeanni, M.J. 1989. Amazonía Peruana ¿Qué Hacer? Interamerican Development Bank, Washington, D.C. Unpublished report.

FAO. 1981. *Los Recursos Forestales de America Tropical.* FAO. Rome.

IUCN. 1988. Peru: Conservation of Biological Diversity. Draft. World Conservation Monitoring Centre, Cambridge, UK.

IUCN. 1990. *1990 United Nations List of National Parks and Protected Areas.* IUCN, Gland, Switzerland and Cambridge, UK.

Ministerio de Agricultura. 1987. *Plan national de acción forestal 1988-2000.* Lima, Peru.

Poore, D., Burgess, P., Palmer, J., Rietbergen, S. and Synott, T. 1989. *No Timber Without Trees: Sustainability in the Tropical Forest.* Earthscan Publications Ltd, London.

Suárez de Freitas, G. 1990. Estudio del Pais: Peru. Lima. Unpublished report prepared for the IUCN Forest Conservation Programme workshop, "Realistic Strategies for Tropical Forest Conservation" in Perth, Australia.

World Resources Institute (WRI). 1990. *World Resources: 1990-91.* Oxford University Press, New York.

PERU
TOTAL FOREST AREA

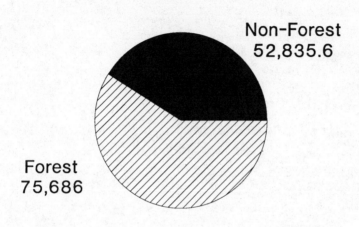

Non-Forest
52,835.6

Forest
75,686

Note: values given in 1000s of ha

PERU
LAND USE DESIGNATIONS

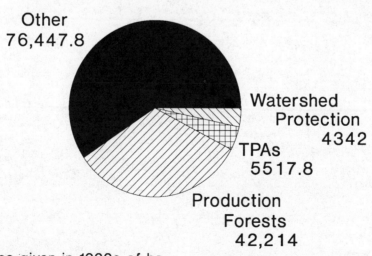

Other
76,447.8

Watershed
Protection
4342

TPAs
5517.8

Production
Forests
42,214

Note: values given in 1000s of ha

TRINIDAD AND TOBAGO

Based on the work of Clarence Bacchus

1. Conclusions and Recommendations

Of all the Latin American ITTO members, Trinidad and Tobago stands alone as possessing the only commercial-scale production forest estate which approaches sustainability. However, the forest management system is tending to reduce or eliminate some commercial tree species. Additional silvicultural treatments are needed in order to ensure the continued production of high value timber.

Much of the current production forest estate was delineated without adequate consideration for the preservation of biological diversity. The totally protected area network is small. The government has not yet approved plans developed over eleven years ago for the creation of a comprehensive parks and reserves system. This proposed system would be established within the production forest estate. The establishment of plantations on degraded lands could make up for much of the lost productive capacity.

2. Extent, Status and Security of TPAs

The government of Trinidad and Tobago retains control of 255,687ha of forest. This represents 83% of the total forest area and half of total national territory. Very little of this has been set aside as totally protected areas. Most of the state forests are managed for multiple-use which includes the preservation of natural plant and animal communities but which also allows for the sustainable exploitation of timber and game resources.

There are currently six protected area units which are greater than 1000ha. Together these total 17,913ha. Numerous smaller protected units are scattered throughout both islands. The six largest protected areas are all classified as Wildlife Sanctuaries. Hunting is prohibited in Wildlife Sanctuaries, but timber extraction and quarrying is sometimes allowed. Even this very small and restricted network cannot, therefore, be properly considered as "totally protected area" as defined by IUCN categories I-V.

The conservation status of the Wildlife Sanctuaries is extremely problematic. For instance, large portions of the Kronsadt Island Wildlife Sanctuary and the Valencia Wildlife Sanctuary have been extensively mined. The Valencia Sanctuary has been heavily logged as well.

In 1980, the Government of Trinidad and Tobago, with support from the Organization of American States, drafted a plan for the establishment of a National Parks and Protected Areas system. The plan recommended the creation of 61 new units representing 13% of total land area. All twenty-six ecological communities found in both islands are represented in the proposed system, although in five units, the remaining vegetation is not sufficiently extensive to constitute viable ecosystems. The government has not yet approved the plan although the expectation is that eventually all or part of the proposed areas will be afforded legal protection.

3. Extent, Status and Security of Production Forests

According to the National Forest Resource Plan, 224,513ha are classified as production forests. Only 165,000ha of this total, however, is deemed capable of sustained wood production. This may change, however, as some of this area will be transferred to the parks and reserves system. Approximately thirteen thousand hectares of the production estate are plantation forests of pine and teak.

In 1989 plantation forests produced 19,400 m^3 and natural forests produced 42,655 m^3. Plantations and natural forests in Trinidad and Tobago are capable of production rates of 15 $m^{3/year}$ and $1.5\text{-}2m3$/year respectively. The conversion of native forest into artificial forests has ceased.

A recent report by the International Institute for Environment and Development (Poore *et al.*, 1989) cited Trinidad and Tobago as possessing the only example of operationally sustainable natural forest management among all the ITTO member countries in Latin America. The report noted the following:

1.   Resident forest guards provide a degree of protection to production areas.

2.   Management objectives are defined.

3.   Most logged areas are covered by working plans (although sometimes these are not up-to-date).

4.   Logging activities are subject to some control.

Based on the above criteria and historical observations, the IIED reviewer determined that logging operations are sufficiently well managed to be considered sustainable. However, there is room for improvement. Management is not intensive and silvicultural treatments are rarely applied. Management prescriptions are not followed strictly.

Timber sales are confined to production blocks and are repeated according to a periodic cutting cycle. In the past, trees to be cut were selected at the discretion of the purchaser. This system was found to be unsatisfactory and currently forest technicians mark trees harvested. Restrictions are imposed on the number of trees removed and the degree of canopy opening. Steep slopes and buffer areas around streams and rivers are not logged.

Little information is available on the effects of logging on species diversity. One of the negative effects of the selective logging system is the reduction of commercial tree species. The effects on non-tree species and plants are less well understood. It is worth noting that in at least two cases, logging activities have been suspended in order to protect rare and endangered species.

The Forestry Division began a continuous inventory of tree species in 1981, which has yielded much useful information. This effort is complemented to some extent by a programme designed to evaluate wildlife habitat and resources which is operated by the Wildlife Section of the Forestry Division. All forest management plans have a wildlife component.

Existing wildlife programmes are oriented towards the sustained exploitation of these resources. Hunting is controlled through the issuing of licenses and controlling the seasons when hunting is permitted. Non-timber and non-game resources receive scant attention in current management plans.

Local conservation organisations and the Forestry Division are polarised on the issue of forest management. It has been difficult for the Forestry Division to meet the conflicting demands for continuous timber supply and ecosystem conservation. The government has taken steps to address environmental concerns by appointing a cabinet-level committee on the environment as well as special committee on biological diversity conservation and by creating the Ministry for the Environment.

Basic Forest Statistics: Trinidad and Tobago

Total Land Area:

513,000ha	(World Resources Institute, 1990. p.268)
512,599ha	(Bacchus and McVorran, 1990. p.1)
512,400ha	(Poore *et al.*, 1989. p. 80)
512,800ha	(World Conservation Monitoring Centre, 1988. p.109)

Total Forest Area:

307,535ha	(Bacchus and McVorran, 1990. p.2)
224,000ha	(WRI, 1990. p.268)
208,000ha	(WRI, 1990. p.292)
272,800ha	(Poore *et al.*, 1990. p.88) (tropical moist forest)

Deforestation Rate:

0.4%	(WRI, 1990. p.292)

Production Forest Estate:

224,513ha	(Bacchus *in litt.*, 1991.) (this includes 13,133ha of pine and teak plantations)
14,000ha	(WRI, 1990. p.292) (managed closed forests)
94,500ha	(Poore *et al.*, 1990. p.80)
99,700ha	(WCMC, 1988. p.110.)
102,300ha	(Oxford Forestry Institute, undated. p.36)

Watershed Protection Forests:

32,200ha	(Poore *et al.*, 1989. p.80)
31,700ha	(WCMC, 1988b. p.110)
37,000ha	(OFI, undated. p.36)

Totally Protected Area:

17,913ha	(derived from: IUCN, 1990 and WCMC, 1988b[1])
19,462ha	(Bacchus and McVorran, 1990. p.4)
96,000ha	(Bacchus *in litt.*, 1991) (proposed total)
16,088ha	(WRI, 1990. p.300)
15,278ha	(IUCN, 1990. p.172)
16,000ha	(WCMC, 1988. p.110) (Wildlife Sanctuaries Only)

Number of Units:

6	(WRI, 1990. p.300)
6	(IUCN, 1990. p.172)
13	(Bacchus and McVorran, 1990. p.4)
61	(Bacchus *in litt.*, 1991.) (proposed total)
13	(WCMC, 1988. p.110)
6	(Derived from: IUCN, 1990 and WCMC, 1988b.)

---

1   This includes all TPAs greater than 1000ha. Numerous small protected areas are not included in this figure.

Bibliography

Bacchus, C. and McVorran, G. 1990. The role of forestry in biological diversity conservation in Trinidad and Tobago. Unpublished report prepared for the IUCN Forest Conservation Programme workshop, "Realistic Strategies for Tropical Forest Conservation" in Perth, Australia.

Bacchus, C. 1991. *In litt.,* 22 February.

Becon, P.R. and Fench, P.R. (Eds), 1972. The wildlife sanctuaries of Trinidad and Tobago. Wildlife Conservation Committee, Ministry of Agriculture, Lands and Fisheries, Port of Spain.

Forestry Division. 1989. *Forestry Resources Policy (1981)*. Government Printery, Port of Spain.

Forestry Division. 1990. The national forest resources plan (Draft). Forest Resources Inventory and Management Section, Forestry Division, Ministry of the Environment and National Service, Trinidad.

Government of Trinidad and Tobago. *The Conservation of Wildlife Act*. Chapter 67:01.

IUCN. 1990. *1990 United Nations List of National Parks and Protected Areas*. IUCN, Gland, Switzerland and Cambridge, UK.

Oxford Forestry Institute. Undated. Productive and Intensively Managed Forests. IIED, London.

Poore, D., Burgess, P., Palmer, J., Rietbergen, S. and Synott, T. 1989. *No Timber Without Trees: Sustainability in the Tropical Forest*. Earthscan Publications Ltd., London.

Thelen, K.D. and Faizool, S. 1980. Plan for a system of national parks and other protected areas in Trinidad and Tobago. Forestry Division/O.A.S. Technical Document. Ministry of Agriculture. Trinidad Syncreators (1970) Ltd.

World Conservation Monitoring Centre (WCMC). 1988. Draft Review of the Protected Area Systems in the Caribbean Islands. Cambridge, UK.

World Resources Institute (WRI). 1990. *World Resources: 1990-91*. Oxford University Press, New York.

TRINIDAD & TOBAGO
TOTAL FOREST AREA

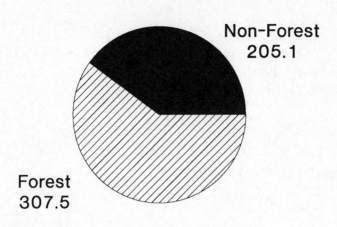

Non-Forest
205.1

Forest
307.5

Note: values given in 1000s of ha

TRINIDAD AND TOBAGO
LAND USE DESIGNATIONS

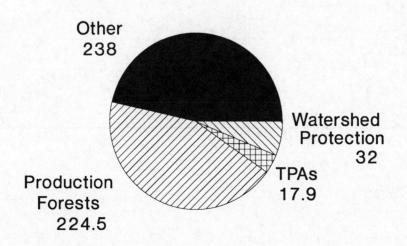

Other
238

Watershed
Protection
32

TPAs
17.9

Production
Forests
224.5

Note: values given in 1000s of ha

LATIN AMERICA OVERVIEW

Based on the work of Gustavo Suarez de Freitas

The regional overview is based on the country reports prepared by the consultants for each Tropical American State Member of the ITTO and on the conclusions and discussions of the Workshop 5: Realistic Strategies for Tropical Forest Conservation, that took place at the 18th IUCN General Assembly, Perth, Australia, 30 November – 1 December 1990.

1. Principle Findings

The following **regional characteristics** were noted:

1) There are over 850 million ha of both dense and open forests in the Latin American region covering 54% of the total land area. This corresponds to almost 22% of the world's forests.

2) The members of ITTO possess 726 million ha of forest. Many people in these countries perceive this as a potentially important resource for their economic development.

3) The contribution of forest industry to economic development is small in comparison with the magnitude of the resource. Most of the countries of the region have a negative import/export balance for forest products which is largely accounted for by deficits in pulp and paper and other finished products. Timber exports are of little relevance to their economies. On average they account for less than 1.3% of export values except for Brazil with 6.71% and Bolivia with 3.34%. In the international tropical timber market, the Latin American participation is minimal (2.2% of the export values).

4) Tropical forests in Latin America are rapidly disappearing. Deforestation estimates vary. Current estimates of deforestation rates for ITTO member countries are as follows:[1]

Bolivia	0.2%
Brazil	1.8%
Colombia	1.7%
Ecuador	2.3%
Honduras	2.3%
Panama	0.9%
Peru	0.4%
Trinidad and Tobago	0.4%

Principal conclusions of the workshop included the following:

1) Tropical forests in Latin America are neither being effectively protected for biological diversity nor are they being managed for permanent production of timber.

2) Well managed permanent production forests hardly exist in Latin America although there are some interesting initiatives in this direction. Logged-over areas tend to be invaded by farmers who clear forests for agriculture.

---

1   World Resources Institute, 1990

Most of the totally protected areas in the region were established less than thirty years ago . The totally protected area systems of the region do not fulfill their objectives because many biogeographic zones are not adequately covered and because few parks and reserves are managed well.

All ITTO member countries in the region possess legally recognised permanent production forests. In each of these countries, however, timber comes mainly from areas which are converted to agricultural use.

Almost all countries have weak forest administrations. Although the laws and regulations are comprehensive in most cases, it has proved difficult to exercise state control over forest resources due to social, economic and political constraints.

All countries in the region suffer from deforestation. This is mainly the result of shifting agriculture and the expansion of grazing areas for livestock. This is the principal threat to the conservation of tropical forests in Latin America. The unplanned and uncontrolled expansion of the agricultural frontier is a result of the lack of economic opportunity in other areas of the economy and the skewed distribution of arable land.

Several countries promote deforestation through legislation which requires forest clearance as a means of establishing legal claim to unoccupied lands. In other cases deforestation is indirectly encouraged through fiscal incentives and subsidies for cattle ranching. Fortunately, some countries are re-evaluating legislation which is both ecologically destructive and economically perverse. Brazil, for instance, has curtailed subsidies to the cattle industry which were responsible for great amounts of deforestation.

Current forest management systems are not, in general, sustainable. A great deal of timber comes from areas destined for non-forest uses and little from those under permanent forest management. Among ITTO member countries, plantations are only significant in Brazil and Colombia.

2. Recommendations

— The boundaries of permanent forest estates must be delimited on the basis of territorial, ecological and economic considerations and with participation of the local people. At the same time totally protected areas, watershed protection forests, extractive reserves and indigenous territories should be defined.

— Every country should set up a permanent system to monitor deforestation accurately.

— All countries of the region should strengthen their forest administration including, not only aspects related to the protection of biodiversity, but also the management of the resource. Currently capacity to manage the forest is inadequate. Forest departments generally do not even have the ability to monitor forest exploitation by private logging concessionaires. Their capacity to enforce existing regulations is very limited.

— Forest legislation should be improved in all the countries so that well managed timber extraction is conducted in legally established permanent production forests.

— There should be incentives for forest conservation and management. The requirement for forest clearance in order to establish tenurial rights should be eliminated. Economic incentives which encourage deforestation should also be eliminated.

— All local people, both native and other rural inhabitants, should benefit from forest management.

— There should be an interactive system in which forest management should be monitored and supported by research. This could ideally be done by promoting cooperation between industry and research centres, including NGOs, with national government collaboration.

— Given that forest management practices oriented to ensure sustainability in the use of the resource give rise to higher costs than the current extraction practices, these costs should be incorporated in the price of the products. This applies to the granting of any sort of incentive, via ITTO, for sustainable forest management. More added value for tropical timber products should accrue to the countries of origin, and these funds should be applied to forest management and reforestation.

3. Present Extent, Status and Distribution of Totally Protected Areas

All the countries of the region have systems comprising several categories of protected natural areas. The names vary from country to country. Table 1 shows the existing categories in each country and their equivalences under the IUCN classification system.

The TPA network coverage, at both the individual country and the regional levels, is not adequate to ensure the conservation of a representative sample of the enormous biodiversity contained in these forests. Percentages for totally protected forests in tropical America fluctuate between 0.1 and 19.5%, with an average for the member countries of the ITTO of 1.7%.

A workshop that took place in Manaus, Brazil in January 1990 identified areas of importance for conservation according to three levels of priority. Map 1 shows available information on the distribution of existing protected areas in Amazonia and Map 2 shows priority areas for conservation.

The list below shows the amount of protected area in each of the ITTO member countries:

Country	Protected Area	# of units	% of land area
Bolivia	6,774,165ha	23	6.2%
Brazil	20,525,324ha	162	2.4%
Colombia	9,016,893ha	42	8.7%
Ecuador	3,007,664ha	15	10.7%
Honduras	709,369ha	34	6.3%
Panama	1,317,382ha	15	17.3%
Peru	5,517,835ha	24	4.3%
Trinidad and Tobago	17,913ha	6	3.5%

These figures clearly show low percentages of totally protected areas; a situation that should be corrected by setting up new national parks and reserves.

Even though protected areas are legally established and given a special management status, implementation is usually lacking. Consequently, there are several management problems common to the countries of the region: land tenure conflicts with neighbouring villages; illegal extraction of fauna and flora, particularly timber; trespassing; and finally, the conflicts that regularly arise with development. These ever-present threats damage the protected areas, diminishing their ability to sustain representative samples of biodiversity. This is due mainly to the weakness of park administrations.

The low status of park administrations in the administrative hierarchy and their weak political support results in meagre budgets that are wholly insufficient to ensure permanent presence and effective control in all the areas. Therefore, it is necessary to strengthen the administration of protected areas by involving the private sector – not only NGOs but also local people – in planning and development.

4. Present Extent, Status and Security of Different Categories of Permanent Production Forests

Various categories of production forests exist in each of the ITTO member countries including: national forests; permanent production forests; watershed protection forests; extractive reserves and communal reserves.

Table 2 gives a list of existing categories by country. Listed below are total areas of production forest for each ITTO member country excluding watershed protection forests:

Country	Production Forests	# of units	% of land area
Bolivia	6,400,000ha	5	5.9%
Brazil	14,761,841ha	42	1.7%
Colombia	1,562,000ha	unknown	1.5%
Ecuador	none	0	0.0%
Honduras	1,642,011ha	4	14.7%
Panama	217,309ha	5	2.9%
Peru	42,214,000ha	43	33.0%
Trinidad and Tobago	224,513ha	unknown	43.8%

Legislation in almost all the countries acknowledges the need for specific areas for forest management for the production of timber and non-timber products. However, most timber is logged from places other than these permanent production forests, notably in Brazil, Bolivia and Peru. The legally established surface for permanent production in Brazil represents a mere 1.73% of the total area of the country.

The total surface of the permanent production forest is insufficient to sustainably satisfy the current demand for timber. Forests cleared for other uses, primarily agriculture and cattle raising, are still the main source of timber for industry.

In Peru, sixty per cent of its forests are part of the permanent production forest network. Of these 5,514,000ha are within five National Forests while 36,700,000ha are within the thirty-eight "Free Availability Forests" ("Bosques de Libre Disponibilidad"). Most timber comes from forests outside these areas. For instance, only 8% of national forests are under concession agreement. A total of 1,454,000ha of forests are leased to concessionaires.

In Bolivia there are only 6.4 million ha of permanent production forests, but, on the other hand, there are concessions covering 22.5 million ha. These cover almost half of the total forest area of the country.

Almost all ITTO member countries in the region have legislation applying to permanent production forests. Others are now developing legal instruments. However, the main obstacle is not legal but rather, poor administrative capacity and the inability to monitor and control timber extraction within permanent production forests.

Even though this legislation is under revision in some countries, like Peru and Bolivia, and could be improved upon in most of them, the poor administrative capacity of the institutions responsible for forest management is the main problem.

Experience has shown that areas designated as permanent extraction forests or forest reserves have been invaded by migratory farmers practising shifting cultivation. This results in the loss of their special status as forest reserves or production forests and their conversion to agricultural or livestock use, with consequent impoverishment and loss of biodiversity.

A good example of this is the current situation of the Peruvian national forests: the 17 national forests existing up to 1975, covering almost 7 million ha have decreased to just 4 with an area of 3.33 million ha. These suffer from encroachment. Furthermore, only 10% of their total area is being managed.

Forest reserves in Brazil have also been fragmented, changed to other management categories, or devoted to other uses. In Colombia, almost all the forests had forest reserve status, but colonisation processes and logging have caused their fragmentation and change in use.

5. Management of the Permanent Production Forests and Implications for Biological Diversity Conservation

Despite legislation in each of the ITTO-member countries providing for the sound management of permanent production forests, in practice, exploitation of forests is not conducted in a sustainable manner.

On the one hand, most of the timber used by the forest industry comes neither from managed natural forests nor from plantations, but from areas to be cleared for the expansion of the agricultural frontier. After selective extraction, these areas are converted to other uses. The permanent production cycle is therefore interrupted.

In Brazil, the permanent production forests produce a negligible amount of roundwood. In Bolivia, there are 22.5 million ha of concessions which cover 50% of all the forests in the country but there are only five officially declared production forests that cover just 6.4 million ha.

In Peru, most of the timber comes either from small contracts (less than 1000ha) without any management plan, or from areas cleared for the expansion of agriculture. Timber generally comes from areas other than the National Forests, where production should concentrate.

Forest production areas, in general, are not secure. They are frequently invaded by migratory farmers who change the land use from forest to agriculture and or cattle raising, disregarding the soil's capacity to sustain such activities.

Therefore, the permanent forest estate does not contribute greatly to biodiversity conservation because forest management does not occur in these areas. Several national forests, forest reserves and other categories have been changed to other uses, generally inappropriate for biodiversity conservation. On the other hand, when forest management does occur in permanent production areas, as in Trinidad and Tobago, it results in forest maintenance and regeneration of the resource, thus contributing to the conservation of biodiversity and ecosystem functioning.

It is important to point out that, even in the best situation, which seems to be the Block system used in Trinidad and Tobago for the sustainable production of *Mora* forests, there is a reduction of biodiversity. Consequently, permanent forest management for biodiversity conservation is important in two ways:

Directly: It reduces the loss of biodiversity, given that both timber and non-timber products are profitably extracted under good management, the forest is maintained without the land being changed to other uses through slash and burn. This is better than not using it for forestry and leaving it open to destruction.

It is necessary to minimise biodiversity losses, although it is accepted that whenever management of a forest resource takes place, and in spite of applying the best available techniques, biodiversity alterations occur. Selection of sites appropriate for management and applications of environmentally sound techniques, as well as silvicultural treatments based on ecological rather than economic criteria are necessary. In addition, it should be noted that sustainable forest extraction should not be carried out on the entire forest surface. Forests should be zoned for strict protection; timber production; non-timber production; watershed protection; and extractive use by local people.

Indirectly: Concentrating forest extraction and management on permanent production areas diminishes pressure on strictly protected areas. The strictly protected areas are vital for the perpetuation of biodiversity. All forest regulations and exploitation should be integrated with the design, establishment and management of these protected areas. Furthermore, permanent production forests concentrate timber extraction and reduce the existing chaos whenever, as in the case of Peru, Ecuador, Brazil, Bolivia and Colombia, most of the timber comes either from areas deforested for reasons other than timber extraction or from forests that are abandoned after valuable woods have been extracted thus leaving them open to invasion or clearing.

The practice of establishing permanent management areas adjoining protected areas, as in the Palcazu Valley, in Peru, favours a lessening of human pressure on the protected area, although it may be more beneficial for some flora and fauna species than for others.

In order for permanent production forest to contribute more effectively to the conservation of biological diversity, it is necessary to institute improvements at several levels, notably:

1) Forest legislation should include permanent production forests thus giving them a legal status that would prevent them from being allocated to other uses.

2) Timber production should take place mainly within permanent production forests. However, production in private or community lands should be possible with adequate regulations. States should not only use legal constraints that include severe penalties, but also incentives, so that forest extraction occurs within permanent production areas and not outside them. The timber industry must get involved with forest management. Industry should supply itself from managed forests, and not be the beneficiary of forest "mining", largely promoted by the industry to lower costs, as has been happening in almost all the countries in the region.

3) Industry, national governments and the international community, through ITTO, should coordinate efforts to meet the higher costs resulting from forest management following more refined biodiversity conservation criteria and ensure a fair compensation for this additional effort. Together with the passing of regulations and guidelines for the best possible forest management, economic and financial incentives should be used to ensure profitability.

4) Decisions on the size and location of a permanent production area should be based on both social and economic parameters as well as ecological ones. It should start from the resource's capacity for sustainable production, taking into account its conservation needs. Locating these areas would depend not only on where industry and markets are located, but also on the location of totally protected areas. Such decisions should be part of any national or regional ecological zoning plan.

5) Local people, industry, small-scale extractors and government agencies must participate in the management of a permanent production forest. Mechanisms vary from country to

country. Rather than granting concessions for long periods, it would be better to invite industry, large-scale extraction companies and organised small-scale loggers, to bid for cutting lots previously inventoried, delimited and prepared for extraction. Loggers would pay for standing wood, and would be required to extract it according to an approved logging plan. They would pay an amount that covers all the pre- and post-logging costs needed to ensure the area's sustainable use. These tasks could be undertaken by the forestry division or by hired contractors. The advantages of this system are various, and some countries of the region are moving toward this kind of management.

6)   Forest agencies or divisions should control extraction activities in permanent production forests in order to contribute to the conservation of biological diversity and ensure reforestation. For instance:

- zone permanent production forests around untouched areas of each forest type;

- protect watercourses;

- use the best extraction techniques to minimise damage caused to residual stands, soil compaction and disturbance caused by road opening and logging trails;

- ensure that trees that are important for feeding, concentration and nesting sites, and that keystone tree species are neither removed nor damaged;

- promote selective extraction instead of clear cutting of extended surfaces;

- promote extended cutting cycles; and

- guard key habitats of certain threatened species.

It is important to point out the need to manage secondary forests. It is possible to make intensive use of second-growth forests not only for timber but for multiple-use. There are interesting examples of secondary growth forests as permanent production areas, based on plantations and management (including regeneration) of fallow forests.

There is a wide range of other resource management categories used in the non-protected areas in the countries members of the ITTO in tropical America; including extractive reserves (Brazil), protection forests and multiple-use areas. In these, a level of harvesting of products other than timber is permitted. These management categories are important for territorial delimitation and, in various circumstances, constitute a buffer for totally protected areas.

6. Ideas for Projects

All the case studies of the member countries of the ITTO agree in their principle recommendations about priority projects for the ITTO. All participants in the workshop, with the exception of the Colombian representative, agreed that labelling and certification of timber from sustainably managed forests was a good idea.

It is recommended that ITTO should assist its members to:

1)   Carry out territorial delimitation (ecological/economic zoning) within all the countries prior to selection of forest areas for permanent management.

2)   Achieve effective management and establishment of a representative network of totally protected natural areas in the tropical forests.

3)   Establish and manage a network of permanent production forests.

4)   Set up national deforestation monitoring systems, with measures for control and punishment of infractions.

5)   Reinforce or modify legislation to ensure harvesting is done within permanent production forests.

6)   Ensure that there is interaction between forest management and research.

7)   Conduct national studies to review the protected natural areas systems (either totally protected or for resource management) in order to know the real status of biological diversity and ecosystem coverage and to ensure the efficiency and integration of the different categories.

8)   Review the policies regarding the use of forest and other competitive resources in the tropics, as well as the legislation and management systems that make sustainable management possible.

9)   Provide broad-based national or sub-national plans for forest resources use.

10)   Demonstrate through specific trials, the feasibility of conducting sustainable forest management of timber and non-timber products following guidelines for best practices. Local communities, industry, NGOs and national governments should participate in these activities.

11)   Develop procedures to include biodiversity as an element in forest inventories and subsequent management plans. This is necessary to measure biodiversity conservation.

12)   Install monitoring systems based on remote sensing and ground truthing to monitor changes in forest cover. Technology and models exist, for example in Brazil, for the establishment of these systems.

13)   Promote the labelling in the international market, of products that come from managed tropical forests. This promotion should be linked with incentives such as better prices, access to "green markets" or any other favourable conditions (e.g. special credits).

14)   Promote the change from a forest industry which uses great volumes or raw material but adds little value to the final product to one which processes smaller volumes more efficiently and adds more value.

15)   Establish a network of projects to demonstrate forest management, at least two for each member country.

16)   Establish a consultant service, primarily through South-to-South cooperation among producing countries, for forest management projects in permanent production forests.

Table 1:	Categories of Natural Protected Areas Existing in Each Tropical American Country Member of ITTO, and its Equivalence under the IUCN System	
COUNTRY	**CATEGORIES**	**IUCN CATEGORIES**
BOLIVIA	National Parks and Wildlife Reserve	II
	National Wildlife Refuge	IV
	National Wildlife Sanctuary	II, IV
	National Hunting Reserve	VIII
	Biological Station	I, II
	Permanent Production Forest	IV
	Protection Forest	VI
	Biosphere Reserves	IX
BRAZIL	National, State and Municipal Parks	II
	National, State and Municipal Biological Reserve	I
	Forest Reserve	VI
	Federal, State and Municipal Ecological Station	I
	Environmental Protection Areas	V
	Biosphere Reserve	IX
COLUMBIA	National Park (Parque Nacional Natural)	II
	Natural Reserve	I
	Unique Natural Area	III
	Flora Sanctuary	IV
	Fauna Sanctuary	IV
	Parkway	V
	Biosphere Reserve	IX
ECUADOR	National Park	II
	Ecological Reserve	I
	Fauna Production Reserve	VIII
	National Protected Area	V
HONDURAS	National Park	II
	Natural Monument	III
	Biosphere Reserve	IX
PERU	National Park	II
	National Reserve	IV, VIII
	National Sanctuary	III
	Historical Sanctuary	III, V
	Protection Forest	VIII
	Reserved Zone	VI, III
	Game Reserve	VIII
	Biosphere Reserve	IX
PANAMA	National Park	II
	Wildlife Refuge	IV
	Recreational Area	V
	Forest Reserve	VI
	Protection Forest	V
TRINIDAD and TOBAGO	Wildlife Sanctuaries	IV
	Nature Reserves	III
	Forest Reserves	VI

Table 2: **Categories of Permanent Production Forest by Countries**

COUNTRY	CATEGORIES
BOLIVIA	Bosque Permanente de Producción (Permanent Production Forest) Bosque de Uso Múltiple (Multiple Use Forest) Bosque Especial (Special Forest)
BRAZIL	Bosques Nacionales (National Forests) Reservas Extractivistas (Extractive Reserves)
COLUMBIA	Reserva Forestal Productora (Production Forest Reserve) Reserva Forestal Productora-Protectora (Production-Protection Forest Reserve)
ECUADOR	
HONDURAS	Bosques Productivos (Productive Forests) Areas de Uso Múltiple (Multiple Use Areas)
PANAMA	Bosque de Producción (Production Forest) Bosque Mixto de Producción (Mixed Production Forest)
PERU	Bosques Nacionales (National Forests) Bosques de Libre Disponibilidad (Free Available Forests)
TRINIDAD and TOBAGO	Forest Reserves

Bibliography

Bacchus, C. and McVorran, G. 1990. The role of forestry in biological diversity conservation in Trinidad and Tobago. Unpublished report prepared for the IUCN Forest Conservation Programme workshop, "Realistic Strategies for Tropical Forest Conservation" in Perth, Australia.

Carrizosa Umaña, J. 1990. Papel de la Actividad Forestal en la Conservación de la Productividad Biologica: Colombia. Unpublished report prepared for the IUCN Forest Conservation Programme workshop, "Realistic Strategies for Forest Conservation" in Perth, Australia.

CDC-Bolivia. 1990. Rol de la Actividad Forestal en la Conservación de la Biodiversidad Biologica. Unpublished report prepared for the IUCN Forest Conservation Programme workshop, "Realistic Strategies for Tropical Forest Conservation" in Perth, Australia.

Flores Rodas, J.G. and Mateo Molina, E. 1990. Study of the role of biodiversity conservation in Honduras. Unpublished report prepared for the IUCN Forest Conservation Programme workshop, "Realistic Strategies for Tropical Forest Conservation" in Perth, Australia.

IUCN. 1990. *1990 United Nations List of National Parks and Protected Areas.* IUCN, Gland, Switzerland and Cambridge, UK.

Poore, D., Burgess, P., Palmer, J., Rietbergen, S. and Synott, T. 1989. *No Timber Without Trees: Sustainability in the Tropical Forest.* Earthscan Publications Ltd, London.

Rodriquez, E. 1990. Untitled. Unpublished report prepared for the IUCN Forest Conservation Programme workshop, "Realistic Strategies for Tropical Forest Conservation" in Perth, Australia.

Schubart, H. 1990. Commercial forestry and the conservation of biodiversity in tropical forests: the situation in Brazil. Unpublished report prepared for the IUCN Forest Conservation Programme workshop, "Realistic Strategies for Tropical Forest Conservation" in Perth, Australia.

Suarez, L. 1990. El papel de la actividad forestal en la conservación de la diversidad biológica del Ecuador. Unpublished report prepared for the IUCN Forest Conservation Programme workshop, "Realistic Strategies for Tropical Forest Conservation" in Perth, Australia.

Suárez de Freitas, G. 1990. Estudio del Pais: Peru. Lima. Unpublished report prepared for the IUCN Forest Conservation Programme workshop, "Realistic Strategies for Tropical Forest Conservation" in Perth, Australia.

World Resources Institute (WRI). 1990. *World Resources: 1990-91.* Oxford University Press, New York.

Map 1: Protected Areas of the Amazon Basin

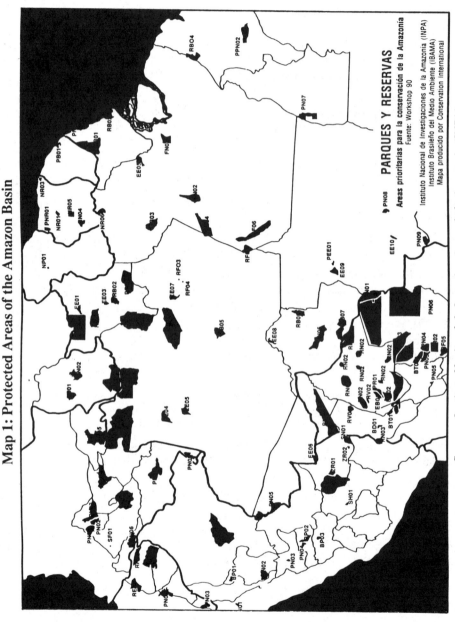

Source: Instituto Nacional de Investigaciones de la Amazonia (INPA)

BRAZIL

Forest Reserves
RF01 Parima
RF02 Rio Negro
RF03 Walter Egler
RF04 Adolfo Duke
RF05 Mundurucania

National Parks
PN01 Cabo Orange
PN02 Pico da Neblina
PN03 Jau
PN04 Amazonia
PN05 Serra do Divisor
PN06 Pacaas Novos
PN07 Araguaia
PN08 Chapada dos Guimarães
PN09 Pantanal Matogrossense

Proposed National Parks
PPN01 Rio Branco
PPN02 Mirador

Ecological Stations
EE01 Maraca
EE02 Maraca-Jipioca
EE03 Caracarai
EE04 Juami-Japurá
EE05 Jari
EE06 Jutai-Solimões
EE07 Anavilhanas
EE08 Cunia
EE09 Iowe Jurema
EE10 Serra das Araras

Proposed Ecological Stations
PEE01 Sema

Biological Reserves
RB01 Mucajai
RB02 Lago Piratuba
RB03 Rio Trombetas
RB04 Gurupi
RB05 Abufari
RB06 Jaru
RB07 Guapore

Proposed Biological Reserves
PRB01 Oiapoque
PRB02 Rio Anauá
PRB03 Jatapu

National Forests
FN01 Caxiuana
FN02 Aveiro

VENEZUELA

National Parks
PN01 Yapacana
PN02 Archipelago los Roques
PN03 Serranía de la Neblina

COLOMBIA

National Parks
PN01 Cordillera de los Picachos
PN02 Tinigua
PN03 Serranía de la Macarena
PN04 Nukak
PN05 Puinawai
PN06 La Paya
PN07 Chiribiquete
PN08 Cahuinari
PN09 Amacayacú

Flora Sanctuaries
SF01 Isla de la Corota

ECUADOR

Ecological Reserves
RE01 Cayambe-Coca

Faunal Production Reserves
RF01 Cuyabeno

National Parks
PN01 Sangay
PN02 Yasuni
PN03 Podocarpus

BOLIVIA

National Parks
PN01 Noel Kempff Mercado
PN02 Isiboro Sécure
PN03 Carrasco Ichilo
PN04 Amboró
PN05 Torotoro
PN06 Santa Cruz la Vieja

Biological Stations
EB01 Beni

Wildlife Reserves
RN01 Manuripi Heath Amazonia
RN02 Lagunas del Beni y Pando
RN03 Ulla Ulla
RN04 Noel Kempff Mercado

Wildlife Refuges
RV01 El Dorado
RV02 Estancias Elsner Espiritu

Permanent Protection Forests
BT01 Bella Vista
BT02 Sajta Ichilo

Forest Reserves
RF01 Itenez
RF02 Rio Boopi
RF03 Covendo
RF04 Chiquitania
RF05 Rio Grande Masicuri

Permanent Production Forests
BD01 Quinera del Aten
BD02 Chimanes
BD03 Guarayos
BD04 Bajo Paragua
BD05 El Chore

Regional Parks
PR01 Yacuma
PR02 El Pirai

PERU

National Parks
PN01 Cuervo
PN02 Rio Abiseo
PN03 Tingo Maria
PN04 Yanachaga-Chemillén
PN05 Manu

Protection Forests
BP01 Alto Mayo
BP02 San Matias-San Carlos
BP03 Pui-Pui

Historical Sanctuaries
SH01 Machu Picchu

National Sanctuaries
SN01 Pampas del Heath

Reserved Zones
ZR01 Manu
ZR02 Tambopata

National Reserves
RN01 Pacaya Samiria

SURINAME

Nature Reserves
NR01 Raleighvallen-Voltzberg
NR02 Wia-Wia
NR03 Wane kreek
NR04 Eilerts de Haan Gebergte
NR05 Tafelberg
NR06 Sipaliwini

Proposed Nature Reserves
PNR01 Kabunkreek

GUYANA

National Parks
NP01 Kaieteur

FRENCH GUIANA

Areas of Biotope Protection
PB01 Région de Kaw

Map 2: Priority Areas for Biodiversity Conservation in Amazonia. Proposed at Workshop 90, Manaus, Brazil, January 10-20, 1990.

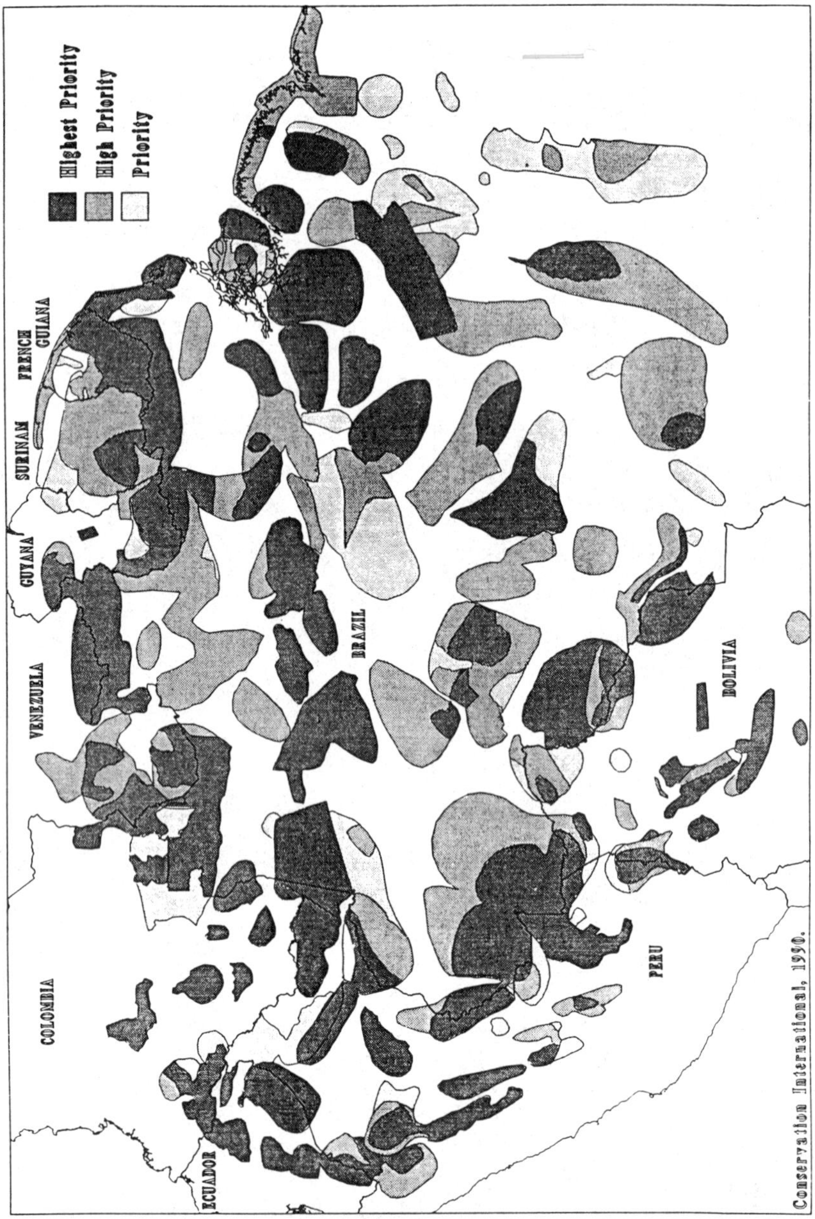

Source: Conservation International. Washington, DC.

COUNTRY STUDIES

AFRICA

CAMEROON

Based on the work of M. Amine and Joseph Bawak Besong

1. Conclusions and Recommendations

In 1981, an attempt was made to reconcile some of the legal inconsistencies in the forest laws of the former Anglophone and Francophone provinces. The resulting legislation, which is the current forestry law, requires that 20% of the surface area of the country be protected as part of the property of the state. However, this same legislation abolished the category of "Forest Reserves" and replaced it with those of "Production" and "Protection Forest". The allocation of the former Forest Reserves to other currently legal categories has not yet been widely implemented. The implications for conservation of this transfer of the reserves to production forests should be kept in mind. Total existing and proposed protected areas cover some 15% of the national territory, or 2.5 million ha short of the legal goal.[1]

The national system of protected areas is both extensive and representative of the diverse biotic communities. Unfortunately, poor standards, inefficient management and a small staff reduce the effectiveness with which this system functions. Improving the system will entail a reconciliation of legal inconsistencies and establishment of national parks in forested areas.

Afromontane forests ought to be better represented in the protected areas system, perhaps by extending the current boundaries of the Bambuko Forest Reserves or Faro National Park and providing protection status for Mont Manengouba and Monte Bakossi. Other areas to consider protecting include marine parks (Rocher du Loup in the Campo Game Reserve), swamp forests (Nyong/Long river) and mangroves.

Fiscal structures of the forest sector ought to be simplified. They are currently over-complex and lead to inefficient exploitation.

2. Extent, Status and Security of TPAs

The protected area system of Cameroon is comprised of national parks and wildlife reserves (administered by the Ministry of Tourism) and the forest reserves (under the Ministry of Agriculture). Protected areas are supposed to be secured by game guards or forest guards. They should have marked boundaries and management plans. In reality, demarcation is poor, protection is weak and management plans are rare. Management plans are required for forest reserves, however, they are rarely put into action. Allocation of funds appears to be the greatest obstacle to achieving an improved protected area system.

Cameroon has seven national parks only one of which, Korup, is in the dense forest zone. This biome is also protected by seven existing wildlife reserves with a total area of 972,995ha. One of these reserves, Dja, is a Biosphere Reserve and is also recognised as a World Heritage Site.

---

1   IUCN has not been able to substantiate some of the forest statistics used in this report which are inconsistent with those given by other authors, i.e. Sayer *et al.* (in press). A summary of forest statistics available from various sources is provided at the end of this chapter.

There are seven current and planned Protection Forests, covering an area of 69,000ha. The proposed Mbam-et-Djerem National Park (which duplicates the proposal to gazette Pangar-Djerem) would protect 353,180ha in the Guinea-Congolian/Sudanian transition zone.

Establishment of the Korup National Park, with support from WWF-UK, is being carried out with recognition of the need for involvement of local peoples in complimentary rural development activities.

The policing of reserves and their marking and patrolling are vital and require adequate financing to back them. The major constraints to the regulation and management of protected areas are:

1)   There is no national system of planning for location of protected areas or for forestry exploitation;

2)   The laws regulating the TPAs are inadequate. For example, buffer zones are treated as national parks. Other problems include the lack of legal recognition of multiple land-use units, inadequate definitions of TPAs, and the failure to establish guidelines for production and protection forest. Finally, there is clearly an inadequate definition of types and extent of intervention permitted in state forest lands. Several legal categories of land protection (e.g. strict nature reserve, sanctuary, state game ranch, etc.) have never been created and others, such as recreation forests, exist only in rudimentary forms. While alteration of legislation will be necessary, improved laws, by themselves, will not secure conservation objectives given the general lack of commitment to allocating greater resources (both financial and human) to the forestry, wildlife and national park sectors.

The Atlantic coastal forests are richer in biodiversity than other forest types. The fact that they are coastal forests has facilitated access by loggers and they are seriously degraded. The Congolese forests are less densely populated but have also been somewhat degraded. Twenty four critical sites for conservation have been identified in the forest zones of Cameroon.

3. Extent, Status and Security of Production Forests

Tropical moist forest covered 17.5 million ha in 1989, compared to 23 million ha of original forest cover. The productive high forest area is estimated to be 16 million ha with an annual loss of roughly 150,000ha or 0.9%. Productive High Forest is equal to the tropical moist forest minus the areas that cannot be logged due to site conditions (e.g. permanently inundated, steep slope) or legal restrictions, such as national parks and protected areas. Cameroon possessed in 1987, 1.3 million ha of Forest Reserves, of which 60% is actually under tropical moist forest cover.[1]

The forests in Cameroon (totalling to 37% of national land area) are rich in diversity. They include semi-deciduous forests, montane, mangroves, Congolese and Biafran forest. Two forest types, the coastal and Congolese, are most prevalent, amounting to 16 million ha, followed by semi-deciduous and transitory types, totalling 4 million ha.

The most critical sites are located in the western montane forests and the coastal forests. The Congolese forests in the southeast are less threatened and are subject to relatively low extraction rates. This may change, however, as these forests have been targeted for increased logging

---

1   Rietbergen, S. 1988. Natural Forest Management for Sustainable Timber Production. International Institute for Environment and Development. Unpublished draft prepared for the International Tropical Timber Organization.

activity. Cameroon is the sixth largest exporter of tropical hardwood in the world, and number three in Africa.

The permanent forest estate is approximately 1,618,565ha; of which production forests total 1,262,117ha or 9.3% of the national forest land. Protection forests total 68,503ha or 0.4 % of national forest land or alternatively stated, 0.1 % of national territory (Gartlan, 1989). The balance of the permanent forest estate is made up of wildlife sanctuaries and national parks.

An extensive management system has been introduced on the SOFIBEL concession covering 80,000ha in Deng-Deng forest, however, Cameroon could benefit from a more efficient logging control system applied nationwide. Three 200,000ha industrial complexes were planned in the south and the south east.

Basic Forest Statistics: Cameroon

Total Land Area:

46,540,000ha   (World Resources Institute, 1990. p.268)
46,540,000ha   (Sayer *et al.*, (in press))

Total Forest Area:

17,500,000ha   (Amine, 1990 (tropical moist forest as of 1989)
24,980,000ha   (WRI, 1990. p.268) ("Forest and Woodland: 1985-87")
23,300,000ha   (WRI, 1990. p.292) ("Extent of Forest and Woodland, 1980s")
15,530,000ha   (Sayer *et al.*, (in press)) ("Rain Forest")
17,900,000ha   (Rietbergen, 1988. p.8) ("Tropical Moist Forest")
17,920,000ha   (FAO, 1988) ("Closed Broadleaved Forest")

Deforestation Rate:

0.9%   (Amine, 1990) ("Productive High Forest")
0.8%   (WRI, 1990. p.292) (1980s)
0.5 to 1.0%   (derived from: Sayer *et al.*, (in press))
0.8%   (derived from: Rietbergen, 1988. p.8)

Production Forest Estate:

1,262,117ha   (Gartlan, 1989)
1,300,000ha   (Rietbergen, 1988. p.14) (includes watershed protection forests)

Watershed Protection Forests:

69,000ha   (Amine, 1990) (current and planned)
68,503ha   (Gartlan, 1989)

Totally Protected Area:

1,702,200ha   (WRI, 1990. p.300) ("All Protected Areas")
2,099,705ha   (IUCN, 1990. p.56)
2,069,100ha   (Sayer *et al.*, (in press))

Number of Units:

12   (WRI, 1990. p.301)
13   (IUCN, 1990. p.56)

Bibliography

Amine, M. and Besong, J. 1990. Untitled. Unpublished report prepared for the IUCN Forest Conservation Programme workshop, "Realistic Strategies for Tropical Forest Conservation" in Perth, Australia.

FAO, 1988. *An Interim Report on the State of Forest Utilization in the Developing Countries*. FO:MISC/88/7. FAO, Rome, Italy. 18pp.

Gartlan, S. 1989. *La Conservation des Ecosystemes forestiers du Cameroun*. IUCN, Gland, Switzerland and Cambridge, UK. 186pp.

IUCN. 1986. *Review of the Protected Areas System in the Afrotropical Realm*. IUCN, Gland, Switzerland and Cambridge, UK. 256pp.

IUCN. 1990. *1990 United Nations List of National Parks and Protected Areas*. IUCN, Gland, Switzerland and Cambridge, UK.

Rietbergen, S. 1988. Natural forest management for sustainable timber production: the Africa region. Unpublished report prepared for IIED and ITTO.

Sayer, J.A., Harcourt, C., and Collins, M.N. (in press). *The Conservation Atlas of Tropical Forests: Africa*. Macmillan Press Ltd., London.

World Resources Institute (WRI). 1990. *World Resources: 1990-91*. Oxford University Press. Oxford. 383pp.

CAMEROON
TOTAL FOREST AREA

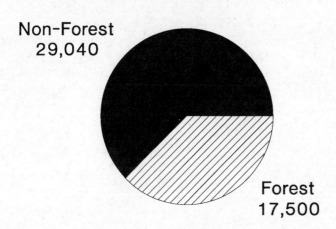

Non-Forest
29,040

Forest
17,500

Note: values given in 1000s of ha

CAMEROON
LAND USE DESIGNATIONS

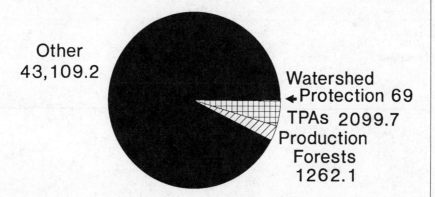

Other
43,109.2

Watershed
◄Protection 69
TPAs 2099.7
Production
Forests
1262.1

Note: values given in 1000s of ha

CONGO

Based on the work of Dominique N'Sosso

1. Conclusions and Recommendations

The existing protected areas of Congo do not cover representative examples of all forest types. In addition, their management is very weak and several of them are subject to agricultural encroachment and heavy poaching. There is an urgent need both to strengthen management of existing areas and to gazette new areas to give comprehensive coverage.

The legal framework for the management of production forests is comprehensive and could adequately provide for the maintenance of biological diversity throughout the permanent forest estate. In the north of the country where logging is highly selective and human population pressure on the residual forest is low, the present timber extraction activities probably have a minimal harmful impact on biological diversity. In the south of the country, and particularly in the coastal area, much forest has been encroached upon after logging, and its biological value seriously degraded.

2. Extent, Status and Security of TPAs

A total of twelve protected areas cover 1,475,100ha.[1] That is 4.3% of national territory. However, this includes 291,000ha in three hunting reserves and the Dimonika Biosphere Reserve which are not protected against human occupation. In addition, significant parts of other areas, which are in theory totally protected, have been encroached upon and seriously degraded. Thus the Lefini Wildlife Reserve, which covers 630,000ha, has lost much of its conservation value due to agricultural activities within its boundaries.

The most valued protected area for forest conservation is the Odzala National Park and the adjacent faunal reserves and hunting reserves. This is now the subject of a major EEC funded project to improve its management and introduce buffer zones.

Some of the existing protected areas would not merit heavy investments in their improved management, particularly as this would run counter to the interest of people living in the area. There is first a need to rationalise the boundaries of these areas, gazette new areas to cover those forest types that are not adequately represented and greatly improve the management capacity of the government institutions responsible for protected areas.

Priority areas for increased protection include parts of the Chaillu Hills and remaining forests in the Mayombe, particularly those closer to the Gabon Frontier than the Dimonika Biosphere Reserve. There are several possibilities for creating new protected areas in the north of the country where population pressures are low. A project is now being launched in the Nouabalé area northeast of Ouésso by Wildlife Conservation International.

---

1   IUCN has not been able to substantiate some of the forest statistics used in this report which are inconsistent with those given by other authors, i.e. IUCN (1990). A summary of forest statistics available from various sources is provided at the end of this chapter.

3. Extent, Status and Security of Production Forests

There are 22,400,000ha of forests remaining in Congo: 17,200,000ha in the north of the country and 5,140,000ha in the south. Of these forests 8,750,000ha are thought to be unproductive for physical and legal reasons, or are under forest fallow. There are 13,690,000ha of productive forests of which 3,360,000ha have been logged, leaving 10,330,000ha still in a more or less intact state. Congo has a comprehensive legal and institutional framework for managing its forest resources and an advanced system of forest management units which should be under integrated, sustained yield management. In general, the prescriptions for the management units concerning exploitation are applied, but those which require that forests be protected and silviculturally treated are neglected. However, the legal basis does exist and could be better applied. In general, the forests which have been logged in the south of the country have suffered degradation as a result of agricultural encroachment, whereas very large areas in the north have been logged lightly for a few high value species, and have subsequently been left to regenerate. It appears that the biological diversity of these northern forests remains largely intact and that they represent one of the better examples worldwide of sustained yield forest utilisation. If roads and infrastructure link these forests to more densely populated areas, their conservation status will be more at risk.

Basic Forest Statistics: Congo

Total Land Area:

34,150,000ha	(World Resources Institute, 1990. p.268)
34,150,000ha	(Sayer *et al.*, (in press))

Total Forest Area:

22,400,000ha	(N'Sosso, 1990)
21,240,000ha	(WRI, 1990. p.268) ("Forest and Woodland: 1985-87")
21,340,000ha	(WRI, 1990. p.292) ("Extent of Forest and Woodland, 1980s")
21,340,000ha	(FAO, 1988) ("Closed Broadleaved Forest")
21,300,000ha	(Rietbergen, 1988. p.8) ("Tropical Moist Forest")

Deforestation Rate:

0.1%	(WRI, 1990. p.292) (1980s)
0.1%	(Derived from: Sayer *et al.*, (in press))

Production Forest Estate:

13,690,000ha	(N'Sosso, 1990) ("Productive Forests" without legal or physical restrictions to logging)

Totally Protected Area:

1,475,100ha	(N'Sosso, 1990)
130,000ha	(WRI, 1990. p.292) ("Protected Closed Forest: 1980s")
1,353,100ha	(WRI, 1990. p.300) ("All Protected Areas")
1,333,100ha	(IUCN, 1990. p.76)

Number of Units:

10	(WRI, 1990. p.300)
10	(IUCN, 1990. p.76)

Bibliography

FAO, 1988. *An Interim Report on the State of Forest Utilization in the Developing Countries.* FO:MISC/88/7. FAO, Rome, Italy. 18pp.

Hecketsweiler, P. 1990. *La Conservation des Esosystemes forestiers du Congo.* IUCN, Gland, Switzerland and Cambridge, UK. 187pp.

IUCN. 1986. *Review of the Protected Areas System in the Afrotropical Realm.* IUCN, Gland, Switzerland and Cambridge, UK. 256pp.

IUCN. 1990. *1990 United Nations List of National Parks and Protected Areas.* IUCN, Gland, Switzerland and Cambridge, UK.

N'Sosso, D. 1990. Untitled. Unpublished report prepared for the IUCN Forest Conservation Programme workshop, "Realistic Strategies for Tropical Forest Conservation" in Perth, Australia.

Rietbergen, S. 1988. Natural forest management for sustainable timber production: the Africa region. Unpublished report prepared for IIED and ITTO.

Sayer, J.A., Harcourt, C., and Collins, M.N. (in press). *The Conservation Atlas of Tropical Forests: Africa.* Macmillan Press Ltd., London.

World Resources Institute (WRI). 1990. *World Resources: 1990-91.* Oxford University Press. Oxford. 383pp.

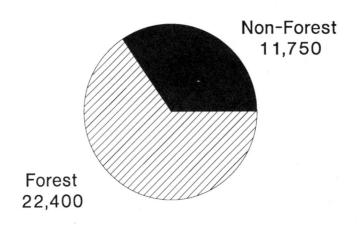

CONGO
TOTAL FOREST AREA

Non-Forest
11,750

Forest
22,400

Note: values given in 1000s of ha

CONGO
LAND USE DESIGNATIONS

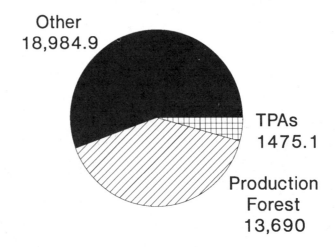

Other
18,984.9

TPAs
1475.1

Production
Forest
13,690

Note: values given in 1000s of ha

CÔTE D'IVOIRE

Prepared by IUCN Staff

1. Conclusions and Recommendations

Côte d'Ivoire has experienced intense deforestation over the past 35 years. Côte d'Ivoire's forests decreased by more than 80% between 1966 and 1990 and they now cover only 3,100,000ha, or 10% of the territory. The realisation of this exhaustion of forestry resources has prompted measures to protect them, including measures which are being developed under a National Forestry Plan.

Total protection areas extend over nearly 6% of the country, covering most types of habitats. In general they are not effectively protected, but international aid has contributed to the development of management plans for the largest areas and help is being provided for the implementation of some of these plans.

Conservation efforts should continue to give priority to the region of the Forest Refuge of Upper Guinea (southwest of the country), particularly the Tai National Park and some areas needing protection near Grabo (Kopé-Haglé Mountains). Improving the management of totally protected areas should be continued, and forestry development activities should be established throughout the Government Permanent Forest Domain.

Forest genetic resource conservation is considered important. In order to achieve this, plantations should include small islands of residual forests. The Tai region's potential for supplying seeds and seedlings should be exploited. The involvement of local communities in forest management should be encouraged and they should be made increasingly aware of agroforestry techniques and generally, of ways to achieve sustainable management of forest resources.

2. The Extent, Status and Security of TPAs

Côte d'Ivoire's system of totally protected areas included eight national parks and two reserves totally 1,917,500ha, or 6% of the national territory. With the addition of the partially protected plant and wildlife reserves, the total rises to 2,019,850ha, or 6.3% of the country. This system covers most types of habitats, from the dense moist forests of the south to the Sudan-savannas with their islands of dry forests in the north. It includes the semi-deciduous forests of the mosaic forest-savanna area in the center of the country and the coastal mangroves.

This system of protected areas suffers from inadequate management and from population pressure. Protection and management are too often ineffective because of a lack of qualified staff and equipment. Poaching is widespread, but competition for various types of land use is the main cause of the rapid degradation of these areas.

International aid is improving the situation: development plans have been prepared for most of the large protected areas, and some of them are already being implemented (i.e. Tai, Comoé, Azagny). In this respect, the Tai National Park is particularly important since it now represents the largest protected area of Guinean forest and contains many endemic plants and animals (Forest Refuge of Upper Guinea). The current project is aimed simultaneously at reinforcing protection of the park and at involving communities in the buffer zone activity. Park managers hope to limit both poaching and deforestation. Several projects have been proposed for extending the system of protected areas to include one near Fresco (mangroves) and another near Grabo (dense forests).

3. Extent, Status and Security of Production Forests

In 1966, forests occupied nearly 17,481,000ha, whereas in 1990 they covered only 3,100,000ha (9.62% of the country.) These forests now represent only 17.7% of those existing in 1966, and most of them are exploited and fragmented. This extensive deforestation results from abusive exploitation of forests and from the agricultural policy of the past 25 years. This policy consisted of encouraging people in overpopulated rural areas to move to less populated forest areas in order to establish new coffee and cocoa plantations. The result was widespread and anarchic deforestation. The rapid decline of the mangroves, on the other hand, resulted principally from the gathering of firewood.

The Government Permanent Forest Domain covers 2,900,000ha; 1,600,000ha of which are in moist forest area (respectively, 9% and 5% of the country). It is made up of 147 forest reserves intended for timber production. The absence of precise limits and of effective protection has caused many reserves to be declassified. The 1987 figures are most telling in this regard when compared with those of 1956 which indicate that, at that time, there were 240 reserves covering 6,000,000ha. In addition, the residual forests of this Permanent Forest Domain are severely fragmented and have been widely exploited, leading to impoverishment in species of commercial value.

In 1988, industrial plantations of exotic and local species covered 66,523ha (0.21% of the country). Attempts have also been made at managing natural or logged-over forests. It is still too early to assess the real value of the methods used, but analysis suggest that investments in forest management yield higher returns than equivalent investments in plantation forestry.

The development of a National Forestry Plan 1988-2015, and international cooperation assistance, should permit initiation of a long-term strategy for the protection and development of forest resources. In particular, the remaining natural or near-natural forests should be surrounded with buffer zones planted with fast-growing species. In order to relieve pressure on the residual forests, establishment is also planned of village plantations in the savanna area, mainly for firewood.

Basic Forest Statistics: Côte d'Ivoire

Total Land Area:

31,800,000ha	(World Resources Institute, 1990. p.268)
31,150,000ha	(Sayer *et al.*, (in press))

Total Forest Area:

6,880,000ha	(WRI, 1990. p.268) ("Forest and Woodland: 1985-87")
9,834,000ha	(WRI, 1990. p.292) ("Extent of Forest and Woodland, 1980s")
2,746,000ha	(Sayer *et al.*, (in press)) ("Rain Forest" as of 1987)
3,100,000ha	(Sayer *et al.*, (in press)) ("Forest and Woodland")
1,300,000ha	(Sayer *et al.*, (in press)) ("Closed Broadleaved Forest")
1,800,000ha	(Sayer *et al.*, (in press)) ("Dry Woodland and Savanna")
2,000,000ha	(Rietbergen, 1988. p.8) ("Tropical Moist Forest")

Deforestation Rate:

5.2%	(WRI, 1990. p.292) (1980s)
6.5%	(derived from: Sayer, (in press))

Production Forest Estate:

2,900,000ha	(IUCN, 1990a)
1000ha	(WRI, 1990. p.292) ("Managed Closed Forests: 1980s")
3,000,000ha	(Rietbergen, 1988. p.14) (Includes watershed protection forests)

Totally Protected Area:

1,917,500ha	(IUCN, 1990a)
2,019,850ha	(IUCN, 1990b. p.77)
648,000ha	(WRI, 1990. p.292) ("Protected Closed Forest: 1980s")
1,958,000ha	(WRI, 1990. p.300) ("All Protected Areas")
1,929,000ha	(Sayer *et al.*, (in press))

Number of Units:

10	(WRI, 1990. p.300)
12	(IUCN, 1990. p.77)

Bibliography

FAO, 1988. *An Interim Report on the State of Forest Utilization in the Developing Countries.* FO:MISC/88/7. FAO, Rome, Italy. 18pp.

IUCN. 1986. *Review of the Protected Areas System in the Afrotropical Realm.* IUCN, Gland, Switzerland and Cambridge, UK. 256pp.

IUCN. 1990a. Untitled. Unpublished report prepared by IUCN staff for the IUCN Forest Conservation Programme workshop, "Realistic Strategies for Tropical Forest Conservation" in Perth, Australia.

IUCN. 1990b. *1990 United Nations List of National Parks and Protected Areas.* IUCN, Gland, Switzerland and Cambridge, UK.

Rietbergen, S. 1988. Natural forest management for sustainable timber production: the Africa region. Unpublished report prepared for IIED and ITTO.

Sayer, J.A., Harcourt, C., and Collins, M.N. (in press). *The Conservation Atlas of Tropical Forests: Africa.* Macmillan Press Ltd., London.

WCMC. 1990. Côte d'Ivoire, managed forest assessment report. WCMC, Cambridge, UK.

World Resources Institute (WRI). 1990. *World Resources: 1990-91.* Oxford University Press. Oxford. 383pp.

COTE d'IVOIRE
TOTAL FOREST AREA

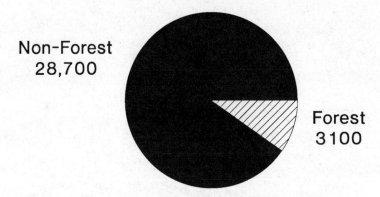

Non-Forest
28,700

Forest
3100

Note: values given in 1000s of ha

COTE d'IVOIRE
LAND USE DESIGNATIONS

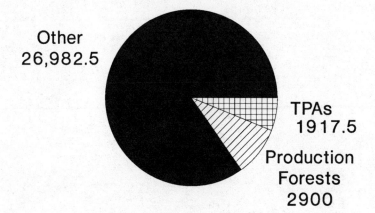

Other
26,982.5

TPAs
1917.5

Production
Forests
2900

Note: values given in 1000s of ha

GABON

Based on the work of Jean Boniface Memvié

1. Conclusions and Recommendations

An extensive system of protected areas exists in Gabon but most of these are available for logging and are therefore not fulfilling their conservation function. Nonetheless, the gazetted areas do cover representative examples of many of the more important forest formations and there is willingness on the part of the government to eventually withdraw them from logging. In some cases, concessions had been awarded before the areas were gazetted and it is difficult to withdraw these concessions without paying heavy compensation to the concessionaires. In the present economic circumstances, the government is unwilling to meet these costs. There is a need for some new protected areas to cover the forest formations in the northeast of the country.

Gabon has a very low human population density and most forests have not suffered encroachment after logging and have regenerated satisfactorily. The density of the principal timber species, *Aucoumea*, is quite high in the coastal zone. This results in logging intensities which tend to be greater than in other parts of Central Africa. Notwithstanding this, the logged over forests are thought to maintain much of their biological diversity. Elephant populations in logged-over forests are higher than those in undisturbed forests.

Little "management" of the forest has been considered necessary since the preferred *Aucoumea* regenerates well when the forest is exploited. Improved communications to the interior, notably the Trans-Gabonese railway line have created greater logging densities in certain parts of the country. This has resulted in greatly increased hunting of wildlife in areas close to the railway but has not in other ways resulted in great loss of forest biodiversity.

2. Extent, Status and Security of TPAs

Gabon has 1,790,016ha of nature reserves. These cover 6.7% of the total national territory. However, it must be stressed that few of these reserves have the legal protection which is normally attributed to national parks or strict nature reserves. Some of them have not been subject to disturbance, notably the central part of the Lopé reserve. Others are unattractive to loggers for physical reasons. In one case, a reserve has been protected as a presidential hunting area. There is an urgent need to develop proper management regimes for the protected areas and to withdraw them from logging concessions. There is also a need to gazette further areas, particularly in the north of the country. A recent study by IUCN proposed an additional fifteen totally protected areas (IUCN 1990).

The Wildlife Division of the Forest Department is responsible for protected area management, but its work is mainly orientated towards anti-poaching activities. It has a reasonable management presence in the Lopé reserve, but is relatively weak in the other existing protected areas. The main forest areas which require additional protection are the Cristal Mountains, the Chaillu Hills, the Gabonese section of the Mayombo, the northeast of the country, the Bélinga area, and the Boka-Boka Mountains.

3. Extent, Status and Security of Production Forests

Gabon has an adequate legal and institutional framework for managing its forests. There is no provision for multiple-use reserves, but there is a system of "Forêts Classées" which are intended for permanent production management. Logging rules and regulations are adequate to ensure a reasonable level of biological diversity in logged over forests. These rules are said to be broadly applied.

In the first zone, nearest the coast, where logging concentrated before construction of the Trans-Gabonese railway, there has been some excessive logging and a decline in the frequency of desirable species. Agricultural encroachment into forests in the coastal zone (and along the railway line) has caused some degradation of the habitat.

In the interior of the country, where population pressure is particularly low, forest exploitation has not been followed by encroachment and regeneration of the desired *Aucoumea* has been satisfactory. The logging concession camps have been focal points for some agricultural encroachment and for very heavy poaching but the impacts of this have so far been relatively localised.

Basic Forest Statistics: Gabon

Total Land Area:

25,767,000ha	(FAO, 1988)
26,766,700ha	(Sayer *et al.*, (in press))
25,767,000ha	(World Resources Institute, 1990. pp.268)

Total Forest Area:

20,500,000ha	(FAO, 1988) ("Closed Broadleaved Forest")
20,000,000ha	(WRI, 1990. p.268) ("Forest and Woodland: 1985-87")
20,500,000ha	(WRI, 1990. p.292) ("Extent of Forest and Woodland, 1980s")
23,544,500ha	(Sayer *et al.*, (in press)) ("Rain Forest")
20,000,000ha	(Rietbergen, 1988. p.8) ("Tropical Moist Forest")

Deforestation Rate:

0.1%	(WRI, 1990. p.292) (1980s)
0.07%	(derived from: Sayer *et al.*, (in press))
0.08%	(Rietbergen, 1988. p.8)

Production Forest Estate:

At present all forests in Gabon are available for logging but significant areas are inoperable because they are seasonally flooded.

Totally Protected Area:

1,790,016ha [1]	(Memvié, 1990)
1,790,000ha [1]	(IUCN, 1990. p.5)
1,753,000 ha	(WRI, 1990. p.300) ("All Protected Areas")

Number of Units:

6	(WRI, 1990. p.300)
5	(IUCN, 1990. p.90)

---

1   Note that according to Sayer *et al.*, (in press) none of the reserves are protected from selective logging.

Bibliography

FAO. 1988. *An Interim Report on the State of Forest Utilization in the Developing Countries.* FO:MISC/88/7. FAO, Rome, Italy. 18pp.

IUCN. 1986. *Review of the Protected Areas System in the Afrotropical Realm.* IUCN, Gland, Switzerland and Cambridge, UK. 256pp.

IUCN. 1990. *1990 United Nations List of National Parks and Protected Areas.* IUCN, Gland, Switzerland and Cambridge, UK.

Memvié, J.B. 1990. Untitled. Unpublished report prepared for the IUCN Forest Conservation Programme workshop, "Realistic Strategies for Tropical Forest Conservation" in Perth, Australia.

Rietbergen, S. 1988. Natural forest management for sustainable timber production: the Africa region. Unpublished report prepared for IIED and ITTO.

Sayer, J.A., Harcourt, C., and Collins, M.N. (in press). *The Conservation Atlas of Tropical Forests: Africa.* Macmillan Press Ltd., London.

Wilks, C. 1990. La Conservation des Ecosystèmes Forestiers du Gabon. IUCN, Gland, Switzerland and Cambridge, UK. 215pp.

World Resources Institute (WRI). 1990. *World Resources: 1990-91.* Oxford University Press. Oxford. 383pp.

GABON
TOTAL FOREST AREA

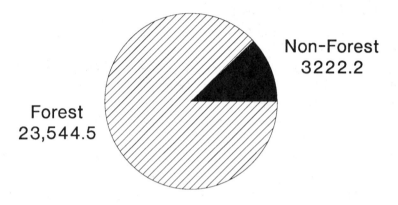

Non-Forest
3222.2

Forest
23,544.5

Note: values given in 1000s of ha

GABON
LAND USE DESIGNATIONS

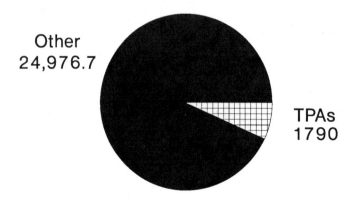

Other
24,976.7

TPAs
1790

Note: values given in 1000s of ha

GHANA

Based on the work of Kwabena Tufour

1. Conclusions and Recommendations

Fifteen percent of Ghana's total land area of 23,853,800ha has been given some level of protection ranging from national parks which are strictly preserved to forest reserves which allow for the extraction of timber. By the year 2000, all forested area outside this network of managed and protected area systems are expected to be converted to other land uses.

Some ecosystem types are at present unrepresented or under-represented. These problems are being addressed, at least in part, by a variety of conservation initiatives undertaken by the Government of Ghana with significant levels of support from the international community. First among these initiatives is the revision of the National Forest Policy.

Timber concession areas could contribute to the preservation of biological diversity while at the same time making an important contribution to national income. However, lack of financial resources, staff and technical ability result in timber areas being degraded.

2. Extent, Status and Security of TPAs

The totally protected area estate of Ghana is administered by the Department of Game and Wildlife and totals 1,074,637ha (4.5% of total land area). This estate comprises three categories which are all managed as strictly protected areas: national parks, strict nature reserves and wildlife sanctuaries. A fourth management category, game production reserves, allows for some exploitation of wildlife and timber resources. Game production reserves total 136,080ha (0.6% of total land area).

Ghana has a long history of conservation initiatives going back to the establishment of the first forest reserves at the turn of the century. The commitment of the government to conservation is reflected in its legislation and in its ongoing efforts to improve natural resource management.

There is an excessive emphasis on conserving useful animals as opposed to all biological diversity. There has been lack of measures to secure public acceptance of protected areas and little provision for conservation beyond the reserve network. It is possible that no forests will remain outside forest reserves or TPAs in Ghana by the year 2000.

The protected areas network excludes some important ecosystem types such as upland evergreen forest and mangrove forests. Semi-deciduous forests are minimally protected. Ninety-one per cent of the wildlife reserve area is located in degraded habitat.

Several new protected areas have been proposed to correct the above-mentioned shortcomings. A revised National Forest Policy will attempt to address conservation problems.

3. Extent, Status and Security of Production Forests

The Forest Department of Ghana has management authority over 2,358,052ha. Management areas are broadly divided into two categories; production forests (1,738,817ha) and protection forests (619,235ha). Portions of both production and protection forests contain large areas of artificial plantations. Protection forest areas serve to protect steep slopes and water catchment areas. Timber extraction in these areas is usually prohibited. Production forests are set aside for sustained production of timber and non-timber resources. Approximately 400,000ha in the Production Forest Estate are located on slopes which are considered too steep to log and are therefore protected.

Significant management problems exist in the forest reserve areas. According to IUCN (1987), "Prospects for conservation of flora and fauna within the forest reserves are not good at present, even though some of these protect important water catchment areas. Apparently, there is little control over what species are felled and the size classes of trees taken". The annual allowable cut is unacceptably high and commercial extinction of several important tree species is likely within 2-3 decades (Gordon 1990).

The Forestry Department is hampered by a lack of adequate funds and manpower shortages. Effective control of concession areas is impaired by the lack of resources.

The Government of Ghana has launched several important initiatives which are aimed at improving the performance of the forestry sector most notably the Forest Inventory Project and the Forestry Resource Management Project. A revised National Forest Policy and a Forest Strategy will be among the products of the latter initiative.

Basic Forest Statistics: Ghana

Total Land Area:

23,002,000ha	(Sayer *et al.*, (in press))
23,002,000ha	(World Resources Institute, 1990. p.268)
23,853,800ha	(Gordon, 1990. p.1)

Total Forest Area:

1,718,000ha	(FAO, 1988) ("Closed Broadleaved Forest")
8,350,000ha	(WRI, 1990 p.268) ("Forest and Woodland: 1985-87")
8,693,000ha	(WRI, 1990. p.292) ("Extent of Forest and Woodland, 1980s")
1,718,000ha	(WRI, 1990. p.292) ("Closed" Forest: 1980s)
8,134,200ha	(Gordon, 1990. p.5) ("Total Forest Area")
1,718,000ha	(Gordon, 1990. p.1) ("Closed Broadleaved")
1,600,000ha	(Rietbergen, 1988. p.8) ("Tropical Moist Forest")

Deforestation Rate:

1.3%	(derived from: Sayer *et al.*, (in press))
0.8%	(WRI, 1990. p.292) (All Forests)
1.3%	(WRI, 1990. p.292) (Closed Forest)
1.3%	(Gordon, 1990. p.1) ("Closed Broadleaved")

Production Forest Estate:

1,738,817ha	(Tufour, 1990)
1,167,000ha	(WRI, 1990. p.292) ("Managed Closed Forests")
1,700,000ha	(Rietbergen, 1988. p.14) (This includes both production and protection forests)

Watershed Protection Forests:

619,235ha	(Tufour, 1990)

Totally Protected Area:

1,074,637ha	(Tufour, 1990)
397,000ha	(WRI, 1990. p.292) ("Protected Closed Forest")
1,175,075ha	(WRI, 1990 p.300) ("All Protected Areas")
1,074,637ha	(IUCN, 1990. p.97)
1,074,637ha	(Gordon, 1990. p.8)
1,311,180ha	(WCMC, 1990. p.6)
1,175,100ha	(WCMC, 1990. p.6) (w/o Game Production Reserves)

Number of Units:

8	(WRI, 1990. p.300)
8	(IUCN, 1990. p.97)

Bibliography

Environmental Protection Council. 1990. Draft environmental action plan. Ghana. Unpublished Report.

FAO. 1988. *An Interim Report on the State of Forest Utilisation in the Developing Countries.* FO:MISC/88/7. FAO, Rome, Italy. 18pp.

Forestry Commission. 1989. Draft national forest policy of Ghana. Unpublished Report.

Forestry Department. 1989. *Ghana Forest Inventory Project Proceedings* 29-30 March, 1989. Ghana. 43pp.

Ghana Forestry Department. Annual reports. Ghana.

Ghartey, K.J.F. 1990. Evolution of forest management in the tropical high forest of Ghana. Unpublished paper presented at West/Central African Rain Forest Conference, Abidjan, 5-9 November, 1990.

Gordon, D.M. 1990. Tropical rain forests: an atlas for conservation: Ghana. World Conservation Monitoring Centre, Cambridge. Unpublished draft.

Hall, J.B. and Swain, M.O. 1978. Distribution and ecology vascular plants a tropical rain forest in Ghana. *Geobotony.*

Hawthorne, W. 1990. *Field Guide to the Forest Trees of Ghana.* Ghana Forestry Series 1. NRI/ODA, London. 278pp.

International Tropical Timber Organization, 1990. *ITTO Guidelines for the Sustainable Management of Natural Tropical Forests.* ITTO, Yokohama, Japan. 18pp.

Irvin, 1960. *Flora of Ghana.* Publisher unknown.

IUCN. 1987. *IUCN Directory of Afrotropical Protected Areas.* IUCN, Gland, Switzerland and Cambridge, UK.

IUCN. 1988. Conservation of Biological Diversity. Unpublished draft. Cambridge, UK.

IUCN. 1990. *1990 United Nations List of National Parks and Protected Areas.* IUCN, Gland, Switzerland and Cambridge, UK.

McNeely, J.A., Miller, K.R. *et al*, 1990. *Conserving the World's Biological Diversity.* World Bank/WRI/IUCN/CI/WWF-US.

Munasinghe, M. and Wells, M., 1990. Protection of biological diversity through local community development. Paper presented at West/Central Africa Rain Forest Conference, Abidjan 5-9 November, 1990.

Poore, D., Burgess, P., Palmer, J., Rietbergen, S. and Synott, T. 1989. *No Timber Without Trees: Sustainability in the Tropical Forest.* Earthscan Publications Ltd, London.

Rietbergen, S. 1988. Natural Forest Management for Sustainable Timber Production: Africa. Unpublished draft prepared for the International Tropical Timber Organization and the International Institute for Environment and Development.

Sayer, J.A., Harcourt, C., and Collins, M.N. (in press). *The Conservation Atlas of Tropical Forests: Africa.* Macmillan Press Ltd., London.

Taylor, C.J., 1960. *Silviculture and Synarcology of Ghana.* Publisher unknown.

Tufour, K. 1990. Status of areas allocated to timber production and their contribution to the conservation of biological diversity. Forestry Commission, Ghana. Unpublished report prepared for the IUCN Forest Conservation Programme workshop, "Realistic Strategies for Tropical Forest Conservation" in Perth, Australia.

Wells, M., Brandon, K., and Hannah, L., 1992. People and parks: linking protected area management with local communities. World Bank, WWF-US, US Agency for International Development, Washington, DC.

World Conservation Monitoring Centre (WCMC). 1990. Protected areas system: Ghana. WCMC, Cambridge, UK. Unpublished draft.

World Resources Institute (WRI). 1990. *World Resources: 1990-91*. Oxford University Press, New York.

GHANA
TOTAL FOREST AREA

Non-Forest
14,867.8

Woodlands
6416.2

Closed
Broadleaved
1718

Note: values given in 1000s of ha

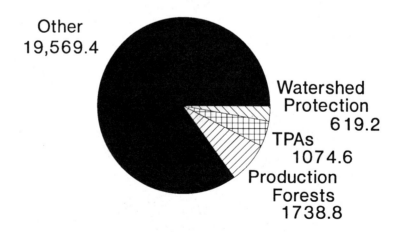

GHANA
LAND USE DESIGNATIONS

Other
19,569.4

Watershed
Protection
619.2

TPAs
1074.6

Production
Forests
1738.8

Note: values given in 1000s of ha

TOGO

Based on the work of O. Nadjombe

1. Conclusions and Recommendations

The combination of intensive human activities and a relatively dry climate has brought about almost total deforestation of the country. Only a few islands of forest remain supporting extremely impoverished plant and animal life. National supply of construction timber production is now very low and far below national consumption. Togo has been an importer of wood for some 20 years.

A Tropical Forestry Action Plan (TFAP) and a Plan of Action for the Environment (PAE) are being prepared. These should strengthen the capacities for forest management. The current priority is to train and deploy sufficient staff to manage the total protection areas and the Permanent Forest Domain. The International Tropical Timber Organization (ITTO) should offer special support for the development and extension of forestry techniques which integrate ecosystem conservation objectives with timber production goals.

The TFAP and the PAE should lead to a national land-use plan taking particular account of forest resources. The notion of conservation of resources must be included among the objectives of management of the natural environment. From an institutional point of view, and for a proper application of this strategy, the present lack of coordination among resource management services must be remedied. The National Commission for Environmental Protection could assume the role of mediator, thereby institutionalising the cooperation and multisectoral planning necessary for the intelligent management of resources.

The few national parks and nature reserves are the best protected areas of the country, but they constitute only the embryo of a system. In developing the system, special attention must be given to local communities. They need to be made aware of conservation needs. Emphasis should be given to agroforestry extension and to the development of village forests. Legislation must be reviewed in order to give clear legal status to the national parks, nature reserves, buffer zones, protection forests, genetic reserves and forests for community use. Forests for production must be created, particularly on the peripheries of total protection areas. Whenever possible, they must include blocks of existing natural forest.

2. Extent, Status and Security of TPAs

Nine wildlife reserves, three national parks and two nature reserves cover a total of 702,850ha[1] (12.4% of the country) and form the system of totally protected areas (81.8%) of the Government Permanent Forest Domain. It should be noted that the parks and nature reserves (415,290ha) do not have an adequate legal base, but are better protected than the nine existing wildlife reserves

---

1   IUCN has not been able to substatiate some of the forest statistics used in this report which are inconsistent with those given by other authors, i.e. Sayer *et al.*, (in press) and IUCN (1990). A summary of forest statistics available from various sources is provided at the end of this chapter.

(287,560ha). "Reserved Forests," (corresponds to Strict Forest Reserves, where all exploitation is forbidden, with the exception of gathering of deadwood and collection of water), are additionally designated as wildlife preservation areas.

Although the parks and reserves are fairly well guarded, they have no management plans. These areas include almost all the major habitats, except for dense, moist, low-altitude forest that is already badly degraded. The other types of forest such as the dense, semi-deciduous forest and woodland and the densely wooded savannas are in no better state of conservation than is the moist forest. This is due to the high population pressures throughout the country. Damage takes the form of poaching, gathering of firewood and deforesting to clear land for agriculture. Frequent fires are accelerating the disappearance of all these wooded expanses. A few islands with the vestiges of sub-montane forests should soon obtain protection status thus completing the present system of protected areas.

3. Extent, Status and Security of Production Forests

The Government Permanent Forest Domain covers 859,500ha (15.2% of the country). Reserved forests, which are intended for timber exploitation, make up 18.2% of this zone (59 reserves totalling 156,650ha). These 859,500ha represent about half of the government's target for conservation and protection of forest resources. They retain only remnants of forests, however, or are covered only by wooded or shrubby savannas. These reserved forests no longer have any value for construction timber exploitation, but only for firewood or timber for local use. Some of them now exist only in name, for they have been completely invaded by local communities. Their value for conservation of biological diversity is thus minimal.

The degradation of the environment by firewood poaching and deforestation for agriculture is continuing and the reserves are no longer effective for conservation. In 1987, the near disappearance of the forests led the Government to suspend all forest exploitation. Forests that are intact now exist only in a very fragmentary and impoverished form, occupying about 5% of the national territory.

At the institutional level, there is a lack of coordination and cooperation among the many agencies responsible for forest resources. The recent preparation of a national TFAP and a PAE should permit some improvement in this area. International assistance has also permitted the initiation of inventories and the development of plantations and the management of natural forests for controlled production of wood. In 1989, industrial plantations, essentially composed of exotic species, covered 20,508ha, or 0.36% of the country, (of which only 9000ha were under proper management).

Basic Forest Statistics: Togo

Total Land Area:

5,439,000ha	(Sayer *et al.*, (in press))
5,439,000ha	(World Resources Institute, 1990. p.268)

Total Forest Area:

304,000ha	(FAO, 1988) ("Closed Broadleaved Forests")
304,000ha	(WRI, 1990. p.292) ("Closed" forest)
1,400,000ha	(WRI, 1990. p.268) ("Forest and Woodland: 1985-87")
1,684,000ha	(WRI, 1990. p.292) ("Extent of Forest and Woodland, 1980s")
1,360,000ha	(Sayer *et al.*, (in press))

Deforestation Rate:

0.7%	(WRI, 1990. p.292) (1980s)
0.6%	(derived from: Sayer *et al.*, (in press))

Production Forest Estate:

156,650ha   (Nadjombe, 1990)

Totally Protected Area:

702,850ha	(Nadjombe, 1990)
463,000ha	(WRI, 1990. p.300) ("All Protected Areas")
646,906ha	(IUCN, 1990. p.171)
647,700ha	(Sayer *et al.*, (in press))

Number of Units:

6	(WRI, 1990. p.300)
11	(IUCN, 1990. p.171)

Bibliography

FAO. 1988. *An Interim Report on the State of Forest Utilisation in the Developing Countries.* FO:MISC/88/7. FAO, Rome, Italy. 18pp.

IUCN. 1986. *Review of the Protected Areas System in the Afrotropical Realm.* IUCN, Gland, Switzerland and Cambridge, UK. 259pp.

IUCN. 1990. *1990 United Nations List of National Parks and Protected Areas.* IUCN, Gland, Switzerland and Cambridge, UK.

Nadjombe, O. 1990. Untitled. Unpublished report prepared for the IUCN Forest Conservation Programme workshop, "Realistic Strategies for Tropical Forest Conservation" in Perth, Australia.

Portas, P. and Sournia, G. 1985. La conservation des ressources naturelles au service du développement socio-economique durable du Togo. Unpublished mission report prepared for IUCN.

Rietbergen, S. 1988. Natural forest management for sustainable timber production: the Africa region. Unpublished report prepared for IIED and ITTO.

Sayer, J.A., Harcourt, C., and Collins, M.N. (in press). *The Conservation Atlas of Tropical Forests: Africa.* Macmillan Press Ltd., London.

World Conservation Monitoring Centre, 1990. Togo managed forest assessment report. WCMC, Cambridge, UK.

World Resources Institute (WRI). 1990. *World Resources: 1990-91.* Oxford University Press. Oxford. 383pp.

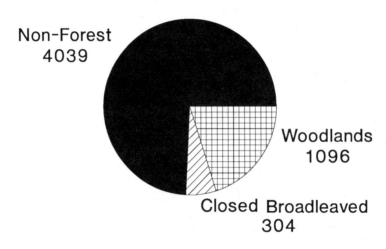

TOGO
TOTAL FOREST AREA

Non-Forest
4039

Woodlands
1096

Closed Broadleaved
304

Note: values given in 1000s of ha

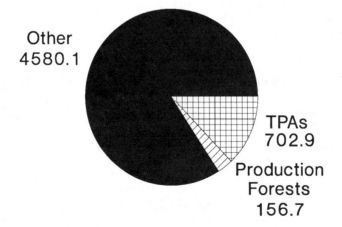

TOGO
LAND USE DESIGNATIONS

Other
4580.1

TPAs
702.9

Production
Forests
156.7

Note: values given in 1000s of ha

ZAIRE

Based on the work of Lumande Kasali

1. Conclusions and Recommendations

Zaire has enormous forestry potential. Regulations governing present exploitation of Zaire's forests, however, do not include specific measures for promoting conservation of the country's exceptional biological diversity. Management plans and strategies recommend zoning and planned management of resources but they have not been implemented. Over-exploitation of forests for timber is one of the dangers facing forests, but gathering of fuelwood and clearing for agriculture have the greater impact. Numerous forest reserves exist, most of them small, but they have been so neglected that many of them are of little value for the conservation of biological diversity.

The extent and management of the national parks system is relatively good. It is administered by Zaire's Institute for Nature Conservation (IZCN), which is the most effective conservation organisation in Central Africa.

The priorities for conservation of forest biological diversity are:

1) Strengthening capacities for management of the existing system of national parks and protected areas.

2) Establishment and strict enforcement of rules for adequate management in forests intended primarily for logging and in multiple-use forests. Strict enforcement of existing management plans should be accomplished as quickly as possible.

3) Selection of sites suitable for conservation of biological diversity with the aim of increasing and complementing the present system of protected areas. The status and zoning of these sites will need to be established in cooperation with all participants and attempts must be made to reconcile written and unwritten law. The areas of priority involve the forests of the mountain chain in the east of the country, the low-and medium-altitude forests of the eastern Cuvette, the mangrove forests (Bas-Zaïre) and the dry forests of Shaba province.

2. Extent, Status and Security of TPAs

The central areas of the Biosphere Reserves and some nature reserves – described in the texts as total protection areas – cover only a tiny area.

The system of totally protected areas consists of seven national parks covering 8,240,000ha[1] or 3.5% of the country. The management and conservation status of these parks vary although as a whole they are better protected than reserved forests (i.e. forest reserves, game preserves, and

---

1 IUCN has not been able to substatiate some of the forest statistics used in this report which are inconsistent with those given by other authors, i.e. Sayer *et al.*, (in press) and IUCN (1990). A summary of forest statistics available from various sources is provided at the end of this chapter.

assimilated reserves). They are the best protected areas of the country, thanks to the efficiency of the organisation in charge of their management, the IZCN.

The national parks include a wide variety of forests: low and medium-altitude dense moist forests; evergreen to semi-deciduous; forests that are swampy or periodically flooded; mountain and transitional forests and; Sudano-Zambezian woodlands. The main gaps in this system are the dense dry Zambezian forests, the mangroves, the swamp forests, those of the eastern Cuvette, and the semi-deciduous forests on the periphery of the Cuvette.

The stated target of Zaire to protect 16% of its territory in national parks, and the projects that are under way (national mangroves park, Okapi National Park, and Mondjo National Park), would meet this target.

3. Extent, Status and Security of Production Forests:

Forests of different types cover about 177,000,000ha, 106,000,000ha of which are dense forests. Aside from national parks, the Government Forest Domain covers about 100,000,000ha (including forests that are reserved but not totally protected).

The area that can be exploited for timber is estimated at nearly 80,000,000ha, but if profitability and accessibility are taken into account, only 30,000,000 to 60,000,000ha are likely to come into production.

In theory, the system of forest reserves (administered by the Directorate of Management of Renewable Natural Resources, DGRNR) and game preserves and related reserves (administered by the IZCN) should improve the conservation of biological diversity of forest ecosystems throughout the country. Many of these reserved forests, however, have been exploited or deforested by the communities living within them. Only a few game preserves (acting as buffer zones in some cases) are useful components of the national parks system on the periphery of the Cuvette. IZCN is denied revenues from game-hunting tourism since hunting was suspended by the government in 1984.

Although they were legally established with the objective of developing forests for production, the forest reserves have fallen short of this goal because almost all of them have been neglected. There are more than 120 of these reserves, totalling more than 517,000ha, but most of them are very small.

Forest clearance for industrial crops is of minor importance compared to shifting agriculture, collection of fuelwood, local timber harvesting, and hunting.

National legislation contains very strict rules for forest exploitation including requirements for management plans, but this legislation is not applied as it should be. Little is done to protect or manage logged-over forests.

Although logging is limited by the difficulties of transporting the timber to export markets, the government's objective is to increase production of construction timber from the present 500,000 m^3 to 1,200,000 m^3 by the year 2000.

Basic Forest Statistics: Zaire

Total Land Area:

226,760,000ha   (World Resources Institute, 1990. p.268)
226,729,000ha   (Sayer *et al.*, (in press))

Total Forest Area:

177,000,000ha   (Kasali, 1990)
105,650,000ha   (FAO, 1988) ("Closed Broadleaved Forest")
105,750,000ha   (WRI, 1990. p.292) ("Closed" forest)
175,630,000ha   (WRI, 1990. p.268) ("Forest and Woodland: 1985-87")
177,590,000ha   (WRI, 1990. p.292) ("Extent of Forest and Woodland, 1980s")
123,200,000ha   (Sayer *et al.*, (in press))

Deforestation Rate:

0.17%   (WRI, 1990. p.292) (1980s)
0.2%   (Sayer *et al.*, (in press))

Production Forest Estate:

80,000,000ha   (Kasali, 1990)

Totally Protected Area:

8,240,000ha   (Kasali, 1990)
5,690,000ha   (WRI, 1990. p.292) ("Protected Closed Forest: 1980s")
8,827,000ha   (WRI, 1990. p.300) ("All Protected Areas")
8,827,000ha   (IUCN, 1990. p.210)

Number of Units:

9   (WRI, 1990. p.300)
9   (IUCN, 1990. p.210)

Bibliography

Doumenge, C. 1990. *La Conservation des Ecosystèmes forestiers du Zaïre*. IUCN, Gland, Switzerland and Cambridge, UK.

FAO. 1988. *An Interim Report on the State of Forest Utilisation in the Developing Countries*. FO:MISC/88/7. FAO, Rome, Italy. 18pp.

Goodson, J. 1988. Conservation and management of tropical forests and biological diversity in Zaire. USAID, Abidjan, Côte d'Ivoire.

Huke, S. and Landu, N. 1988. L'état de conservation des forêts et ses besoins immediats. Etude institutionelle du secteur forestier, Août 1987 – Avril 1988. Département des Affaires Foncières, Environnement et Conservation de la Nature, Kinshasa, Zaire et IIED, Washington, DC.

IIED, 1988. Rapport du séminaire sur la politique forestière au Zaire, Kinshjasa, 11 au 13 mai 1988. IIED, Washington, DC.

IUCN. 1986. *Review of the Protected Areas System in the Afrotropical Realm*. IUCN, Gland, Switzerland and Cambridge, UK. 256pp.

IUCN. 1990. *1990 United Nations List of National Parks and Protected Areas*. IUCN, Gland, Switzerland and Cambridge, UK.

Kasali, L. 1990. Untitled. Unpublished report prepared for the IUCN Forest Conservation Programme workshop, "Realistic Strategies for Tropical Forest Conservation" in Perth, Australia.

Rietbergen, S. 1988. Natural forest management for sustainable timber production: the Africa region. Unpublished report prepared for IIED and ITTO.

Sayer, J.A., Harcourt, C., and Collins, M.N. (in press). *The Conservation Atlas of Tropical Forests: Africa*. Macmillan Press Ltd., London.

World Resources Institute (WRI). 1990. *World Resources: 1990-91*. Oxford University Press. Oxford. 383pp.

ZAIRE
TOTAL FOREST AREA

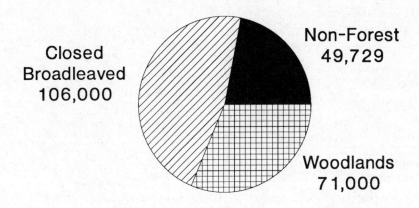

Non-Forest
49,729

Closed
Broadleaved
106,000

Woodlands
71,000

Note: values given in 1000s of ha

ZAIRE
LAND USE DESIGNATIONS

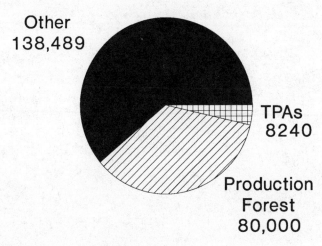

Other
138,489

TPAs
8240

Production
Forest
80,000

Note: values given in 1000s of ha

AFRICA OVERVIEW

Based on the work of
Joseph B. Besong and Francois Wencelius

1. Introduction

This paper presents an overview of country reviews prepared by representatives of African country members of ITTO. These reviews were presented during a regional session of the workshop on "Realistic Strategies for Tropical Forest Conservation" which took place during the IUCN General Assembly in Perth, Australia in December 1990.

The overall objective of the workshop was to examine the role of production forests in the conservation of biodiversity in tropical moist forests (hence referred to as TMF). The objectives of the country reviews were to examine the extent to which the biological diversity conservation function of totally protected areas is complemented by forest areas allocated to various productive uses and to make recommendations on the possible role of ITTO to influence policies and programmes relating to timber production in so far as these impact on biodiversity.

The information used to prepare this paper was provided by country reviews prepared for Cameroon, Congo, Gabon, Ghana, Togo, and Zaire, and by complementary documents provided by IUCN regarding these countries and Côte d'Ivoire.

After a statement of conclusions and recommendations, this overview will discuss the present status of biodiversity conservation and management of TMF resources in the countries under review and then propose how to achieve biodiversity conservation, what strategy should be adopted to reach these goals and how ITTO could possibly contribute in developing this strategy.

2. Principal Conclusions and Recommendations

Present Status of Biodiversity Conservation

In the countries under review, TMFs are facing a tremendous pressure from deforestation and this is the main cause for loss of biodiversity. Remaining TMFs still cover significant areas in Central African countries but are now reduced to small patches in West Africa. Biodiversity in remaining TMF has decreased, but remains high.

A commendable effort to establish protected areas in TMFs has been undertaken by most of the countries under review. However, the existing systems of protected areas are not extensive enough and are often poorly managed.

Little has been done so far to preserve biodiversity in forests outside protected areas. In most of the countries, the permanent multi-purpose forest estate is still small and hardly even managed for timber production with only recent consideration for biodiversity. Most of the TMFs are not gazetted and not protected or managed.

The actual impact of logging on biodiversity depends on the intensity of logging systems. The impact of hunting for bushmeat also depends on human population densities. These impacts are not yet quantified.

Biodiversity conservation is affected by political, legal, fiscal, and institutional issues which call for more commitment on the part of governments and also for policy reforms.

How to Achieve Biodiversity Conservation in TMFs

The only realistic solution to achieve biodiversity conservation in TMFs is to develop a system of extensive protected areas representative of all forest ecosystems and to link it with a system of permanent forests managed for multiple purposes. These actions will be feasible only if they are integrated in a broader approach to solving the land and resource use conflicts affecting biodiversity conservation.

Developing the system of protected areas means strengthening protection and management of existing areas and extending the system to new areas. Developing a multi-purpose permanent forest estate means assigning a large portion of TMFs to gazetted forests which will be managed for production and protection of the environment (soils, water, and microclimate) with due consideration for biodiversity conservation.

The main issues regarding biodiversity conservation are closely related to the conflicts between conservation and other uses of TMF resources. To solve these two conflicts, governments have the central responsibility of developing policy reforms and systems of incentives and will have to share the responsibility of implementing conservation and management activities with local populations, NGOs, producers groups and the private sector.

Proposed Strategy

The strategy to achieve biodiversity conservation in TMF should combine the following:

1) Developing a national and local land-use planning policy in areas covered by TMF.

2) Implementing policy reforms.

3) Carrying out training and research activities.

4) Strengthening field operations.

A land-use planning policy should include a national vision of the best use of the resources of TMF areas that will guide sectoral policies and operational mechanisms to match this vision with local needs and aspirations. All natural resource sectors must be involved as well as sectoral institutions concerned with international trade, debt and general economic policy.

Policy reforms are crucial to improve the legal, fiscal and institutional environment of the future effort to preserve and manage TMFs. Fiscal reforms and new incentives are essential to finance this effort.

A tremendous effort is needed to train and educate local populations, government agents, private entrepreneurs and decision makers in matters related to biodiversity conservation in TMFs. An equal effort is needed to strengthen research.

Field operations should be developed in two successive phases. A first phase would start developing urgent actions to protect critical sites and pilot operations to manage multi-purpose gazetted forests. Simultaneously, land-use planning, policy reforms, and studies would be developed. A second phase would develop the systems of extensive protected areas and permanent forests managed for multiple purposes on the basis of the results of the land-use planning exercise. This would be conducted in an improved legal, fiscal, and institutional environment.

Possible Role of ITTO

ITTO could influence policies and programmes relating to timber production and biodiversity conservation, stimulate an international debate on relevant issues, support the design and adoption of policy reforms and incentives, conduct studies, coordinate pilot operations and provide technical assistance.

3. Present Status of Biodiversity Conservation and Management of TMF Resources

TMFs in the countries under review still contain a great deal of the biodiversity that prevailed before human disturbance became significant. However, biodiversity is under a growing threat in many of these countries and is being depleted in some of them because of explosive deforestation. Measures have been taken to preserve biodiversity by establishing a system of protected areas. However, this system is not efficient and is not yet complemented by a well-run system of multi-purpose permanent forests.

Assessing the status of biodiversity conservation in the countries under review means assessing the following:

1) The extent and location vis-à-vis biodiversity "hot spots" of remaining TMFs.

2) The status of protected areas in remaining TMFs.

3) The status, use and management (if any) of forests outside protected areas in remaining TMFs and their implication for biodiversity conservation.

4) The main policy issues affecting biodiversity conservation.

Extent of TMFs

In the countries under review, TMFs are facing a tremendous aggression from deforestation which is the main cause for loss of biodiversity. Remaining TMFs still cover significant areas in Central African countries, but are now reduced to fragments in West Africa. Biodiversity in remaining TMF has decreased, but remains still high.

The actual pace of deforestation in TMFs varies from one country to an other. Accurate data do not yet exist. Existing figures given by the table in Table 1 are estimates of the situation by the FAO in 1980. The annual deforestation rate ranged from 15,000ha in Gabon to 290,000ha in Côte d'Ivoire. According to preliminary results from more accurate estimate being made by FAO, deforestation has accelerated in the 1980s in countries which possess large areas of TMF, but deforestation rates have decreased in countries where little TMF remains (e.g. Côte d'Ivoire and Ghana).

The two main direct causes of deforestation in TMFs are shifting cultivation and conversion of the natural forest cover into perennial cash crops under individual farms or large scale industrial schemes. In some countries under review (e.g. Côte d'Ivoire) the first two causes are exacerbated by a land-grabbing process which consumes more forest land than is needed by agriculture. These direct causes have their origin in the tremendous growth and movement of population in different countries. As a result of these direct causes, the deforestation process is geographically channelled and facilitated by the development of transportation infrastructure and also by logging activities in TMFs. Logging has no direct influence on the rate of deforestation but it opens roads to farmers and accelerates the deforestation process.

The present estimated TMF coverage, as given in the country reports, is shown in Table 1. This coverage varies from almost zero in Togo to slightly more than 100 million ha in Zaire. TMF cover ranges from almost 0% of the total area of Togo to less than 10% in Côte d'Ivoire and Ghana, more than 40% in Cameroon and Zaire, 65% in Congo, and 86% in Gabon.

TMF still remains in most of the areas where biodiversity is very high but is under various degrees of threat. According to the theory of "forest refugia", there are three biodiversity "hot spots" in the countries under review. First, southern Côte d'Ivoire and Ghana, where biodiversity has decreased since the remaining TMFs have been fragmented by agriculture encroachment and degraded by logging. Second among these are western Cameroon, Gabon and Congo, where biodiversity is at risk. Cameroon suffers from problems similar to those of Côte d'Ivoire and Ghana. In Gabon and Congo, pressure from agriculture and impact of logging is still low. Third, the eastern fringe of the dense forest zone in Zaire, where biodiversity is at risk because of roads from the "dorsale du Kivu" to the river Zaire and threats from mining, logging and agriculture.

Status of Protected Areas

A commendable effort to establish protected areas in TMFs has been undertaken by most of the countries under review. However, the existing systems of protected areas are not extensive enough and are often poorly managed. Protected areas are all gazetted as state forests, but are given many different types of legal status and are under various levels of actual protection. These areas will be considered under two main groups: totally protected areas and wildlife reserves. Specifics of each group and common issues regarding biodiversity conservation will be presented hereafter.

Totally Protected Areas (TPAs): In TPAs, existing legislation provides for total protection. Real on-the-ground protection is most effective in these areas. Indeed, existing TPAs appear to afford the greatest security against encroachment from surrounding populations. There are, however, some variations from one area to an other. There are sixteen TPAs in the TMFs in the countries under review; they include nine national parks (Korup in Cameroon, Azagny and Marahoué in Côte d'Ivoire, Nini-Suhien and Bia in Ghana, Kahuzi-Biega, Maiko, Salonga, and Virunga in Zaire), three Biosphere Reserves (Dimonika in Congo, Luki and Yangambi in Zaire), three areas which are simultaneously Biosphere Reserves and national parks or wildlife reserves (Dja in Cameroon, Odzala in Congo and Tai in Côte d'Ivoire) and one integral natural reserve (Ipassa in Gabon). Most of these TPAs are former forest reserves established during the colonial time. Their TPAs status was conferred recently (in the early 1970s in Côte d'Ivoire, Gabon, Ghana, and Zaire; in the 1980s in Cameroon and Congo).

The area covered by these TPAs (about 8 million ha), is still low compared to the total TMF area in the countries under review (about 175 million ha). TPA cover ranges from almost 0% of TMF area in Gabon, Ghana and Togo to 1% in Congo, about 4% in Cameroon, 6 % in Zaire and 20% in Côte d'Ivoire. The latter ranks high because of the small TMF area that is left.

Wildlife Reserves (WRs): Existing legislation mandates different levels of protection for various categories of WRs. In "Réserves de Faune" in Cameroon, Congo, and Côte d'Ivoire and wildlife sanctuaries in Ghana, extractive activities are prohibited. In most cases, however, hunting is allowed in WRs. This is true for the "Aires d'Exploitation Rationnelle de la Faune" in Gabon the "Domaines de Chasse" in Congo and Zaire and the game production reserves in Ghana. Many of these WRs are subject to encroachment and poaching, and in many countries reviewed (e.g. Cameroon, Congo, and Zaire) some WRs are said to exist only on paper, having been totally invaded and taken over for farming and other uses. In most existing WRs, logging is legally authorised (e.g. Côte d'Ivoire and Gabon) or permitted "de facto" by governments (e.g. Cameroon).

The area covered by existing WRs is about 5 million ha. WRs cover 0% of TMF area in Togo, about 2% in Cameroon, Congo and Zaire, about 5% in Côte d'Ivoire and Ghana, and about 8% in Gabon.

Issues Common to TPAs and WRs: The first issue regarding biodiversity conservation is that TPAs and WRs do not cover the whole range of forest ecosystems and wildlife species that need protection. As documented by the IUCN regional report for conservation and rational utilisation of forest ecosystems in Central Africa (IUCN,1989), existing protected areas cover about 40% of the sites identified as critical for biodiversity conservation in Cameroon, Congo, Gabon, and Zaire. Based on country reviews, Côte d'Ivoire and Ghana also need to widen the range of protected sites, though the scope for this is reduced compared to other countries because of the small area of TMFs left.

The second main issue is that TPAs and WRs are in fact often poorly protected because they lack management plans, real participation of local populations, and guard forces. Management plans are just being developed in a few protected areas under the aegis of IUCN. The buffer zone concept is not taken into account in many existing laws: it is simply omitted (e.g. Congo and Zaire) or misconceived (e.g. Cameroon, where provisions regarding buffer zones are identical to those regarding core protected areas). However, buffer zones are actually envisaged or established for some National Parks in Cameroon, Congo and Zaire, regardless of legal provisions. Local populations around the protected areas and forest dwellers have not been involved in their protection and management. Repressive behaviour on the part of government agents has often led to the adoption of negative attitudes towards the existence of protected areas. The guard forces in charge of these protected areas are understaffed and ill-trained and lack both equipment and financing for efficient operations.

Status of Forests outside Protected Areas

Almost nothing is done in forests outside protected areas to favour biodiversity conservation. In most of the countries, the permanent multi-purpose forest estate (i.e. gazetted state forests other than TPAs and WRs) is still small and hardly managed for timber production with only recent consideration for biodiversity. Most of the TMFs still belong to the so called "protected domain" ("Domaine Protégé" in francophone countries) which is not gazetted and not protected nor managed. The actual impact of logging on biodiversity depends on the intensity of logging systems. The impact of hunting for bushmeat depends on human population densities. Specifics of multi-purpose gazetted forests and "Domaine Protégé" and more general considerations regarding the impact of logging on TMF are discussed below.

Gazetted Forests (GFs): These are gazetted forests other than protected areas. They have been established in TMFs of all the countries under review. The legal status of these GFs provides that they are inalienable, that traditional rights and uses are regulated and that agriculture is not permitted. However, there are some exceptions to this varying from one country to another.

Logging is the direct responsibility of the forestry administration in Cameroon, Congo, Gabon, and Zaire. In Côte d'Ivoire and Ghana, private concessionaires log forests.

These GFs cover a significant area (totalling about 5 million ha) only in countries where TMFs have been (Côte d'Ivoire and Ghana), or are being (Cameroon), dramatically reduced by deforestation. GFs represent 85% of the remaining TMFs in Ghana. In Côte d'Ivoire 60% of remaining TMFs are located in GFs. With the exception of few large GFs, GFs in Cameroon are small and many of them are subject to heavy encroachment. These GFs are mainly production forests. With the exception of Ghana, the criterion of protection of soils and water has seldom been taken into account in gazetting these forests.

There are examples of TPAs (i.e. Tai in Côte d'Ivoire and Korup in Cameroon) being buffered by large GFs. However, most of the latter were not established specifically to complement and enhance the role of TPAs and WRs in preserving biodiversity.

GFs are marginal in Congo, Gabon, and Zaire (where they cover a total of about 0.5 million ha). GFs once established in Congo for protection purposes were degazetted in the late 1970s. GFs in Gabon are limited to artificial plantations and many of the GFs established in Zaire are said to exist only on paper.

Under existing legislation, inventories have to be carried out and management plans have to be developed in GFs. This has hardly started in most of the countries under review. Management plans of GFs are being developed in Ghana on the basis of traditional forest inventories. These plans are aimed at sustained production of timber and non-timber products, but also take into account soil and water protection concerns. However, little has been done to take biological values of forests into account and to consider biodiversity conservation. Management plans of GFs will start being developed soon in Côte d'Ivoire under its recent forestry sector programme. These plans are designed to take biodiversity conservation into account. Successive management plans have been developed in Cameroon for a few GFs (e.g. Deng-Deng), but none has ever been approved and adopted. These plans did not take biodiversity conservation into account. Little has been done to involve local populations in developing and implementing management plans.

"Domaine Protégé" (DP): Forests that are not gazetted (i.e. other than TPAs, WRs, and GFs) belong to the DP, which is still the property of the state. In the DP there are no legal limitations on traditional rights and uses. There are also no actual restrictions on encroachment by agriculture, even if existing legislation contain regulations which prohibit this activity.

Commercial exploitation of timber is regulated by legislation. Authorisation of logging is the responsibility of the state and sometimes also of traditional authorities (e.g. in Zaire and Ghana). The DP is in fact a domain where state legislation on forest resources and traditional rights on land use are in effect simultaneously. This sometimes creates conflicts or at the very least confusion.

TMFs in the DP cover a total of about 160 million ha, almost entirely in Cameroon, Congo, Gabon, and Zaire. This represents 90% of the total TMF area in the countries under review. The DP is where most of the logging takes place in these four countries. Logging is mainly done by private companies with the exception of Congo, where some parastatals also operate in the DP.

Forest resources in the DP are not managed and regulations applying to logging include little consideration of technical forestry aspects or of biodiversity conservation. Regulations are often poorly enforced. Logging concessions are not established according to a clear strategy. There is, however, an exception in Congo where concessions coincide with exploitation units ("Unités Forestières d'Exploitation") determined within so-called management units ("Unités Forestières d'Aménagement") on the basis of a forest inventory. The maximum annual cut is determined for each exploitation unit. Concessions are often issued for a short period (e.g. five years in Cameroon and Côte d'Ivoire, seven years in Congo) which may increase (e.g. up to twenty to thirty years in Congo and Zaire) if the concessionaire invests significantly in wood processing. Short concession periods discourage the concessionaire from caring for the forest. Minimum exploitable diameters are determined by law for each species. A comprehensive inventory and mapping of all trees to be felled is required but real planning of felling is not always carried out. Existing legislation also provides a list of species that cannot be felled because they are endangered species or because they should be preserved for traditional uses. Setting aside patches of critical ecosystems for conservation is possible under existing legislation but this has not yet been widely applied. There are no precise regulations to limit felling damage or to protect forests after felling.

The forestry administrations in charge of enforcing these regulations are often understaffed, poorly trained and lack both equipment and financing for efficient operations.

Impact of Logging, Silvicultural Treatments, and Hunting: The impact of logging on TMFs is only partially documented and the impact on biodiversity in the long run is not yet well known. The same applies to the impact of silvicultural treatments and of hunting.

The impact of logging on the canopy and standing volume of TMF has been documented. Studies carried out in the early 1980s show that from 5% to 9% of the canopy is destroyed by logging depending on the volume felled and the methods of exploitation. These results have been confirmed by a recent study conducted in Eastern Cameroon and preliminary results of another study being conducted in Northern Congo.

Regeneration of forests after the extensive type of logging now practised in most of the TMFs in Central Africa is very close to the natural regeneration process that starts in undisturbed forests each time an old tree falls and opens the canopy. The impact of this type of logging on plant biodiversity is likely to be very low. The regeneration process after more intensive types of logging (like those practiced in Côte d'Ivoire, Ghana, littoral forests in Cameroon, and Mayumbe forests in Congo and Zaire) are certainly different from those in undisturbed forests. The impact of these regeneration processes on biodiversity is not known.

The impact of logging on wildlife is being documented. Forests degraded by logging are said to provide an improved habitat for some species (e.g. gorillas in Cameroon; bongos in Ghana), but intensive logging has proven detrimental to species living in tall trees (e.g. Colobus monkeys in Côte d'Ivoire).

The impact of silvicultural treatments on biodiversity is not yet well known. Experimental silvicultural treatments tested during colonial time in several countries (e.g. Côte d'Ivoire, Ghana, and Zaire) have not been monitored. Silvicultural treatments meant to stimulate the growth of standing trees of commercial species have been successfully tested in Côte d'Ivoire since the early 1980s. These treatments stimulate the regeneration of these species. The impact of these operations on biodiversity has, however, not yet been assessed.

The impact of hunting on biodiversity is not yet quantified but is presumably very serious. Significant depletion of wildlife to supply urban centers with bushmeat has been reported (e.g. in Congo) to affect large areas.

Policy Issues

Biodiversity conservation is affected by political, legal, fiscal, and institutional issues, which call for more commitment on the part of governments and also for policy reforms.

There is not yet a strong commitment to biodiversity conservation in TMFs at the highest political level in the countries under review. National forestry sector policies developed under the Tropical Forestry Action Plan (TFAP) in the 1980s (i.e. Cameroon, Côte d'Ivoire, Ghana) do not give high priority to biodiversity conservation. However, recent (i.e. Zaire) and ongoing (i.e. Congo and Gabon) TFAP exercises give more attention to biodiversity. An Environment Action Plan is being developed in Togo which will give high priority to biodiversity. TFAP should still be considered as the main tool to coordinate governments and donors at the national level in the forestry sector. TFAP, however, needs to be dramatically improved to better take into account biodiversity conservation and participation at the local level.

Existing legislation needs to be improved (e.g. Cameroon and Congo) or completely renewed when outdated (e.g. Zaire) to better regulate biodiversity conservation and protection of the environment. Improvements should mainly deal with status of protected areas and buffer zones, rights of forest dwellers, participation of local populations in conservation and management of forest resources and regulation of in-forest management practices.

The main fiscal issue is that the present levels of forestry taxes that pertain mainly to logging and hunting do not reflect the actual value of these resources. Another issue is that collection of forest revenues is poor and procedures to make local populations benefit from a share of these revenues do not work (e.g. Cameroon and Côte d'Ivoire) or simply do not exist. Forestry fiscal systems need to be improved.

Existing institutional arrangements also need considerable improvement to ensure a more effective involvement of governments, local populations, NGOs and the private sector in conservation and management of TMF resources. Country reviews also stressed the need for more effective enforcement of legislation by courts, better integration of government administrations in charge of forest management and biodiversity conservation (e.g. Cameroon and Ghana) and strengthening of forest services.

4. How to Achieve Biodiversity Conservation in TMFs

The only realistic method of achieving biodiversity conservation in TMF is to develop a system of extensive protected areas representative of all forest ecosystems and to link it with a system of permanent forests managed for multiple purposes. Indeed, the extent of protected areas will be limited by other competing uses of forest resources, mainly land for agriculture, fuelwood and non-timber forest products for households and timber for wood industries and exports. In many countries, the extent of protected areas will never be sufficient to ensure conservation of all species. It is therefore necessary to include forests managed for production and/or protection of the environment in the biodiversity conservation process. These forests provide habitats for the majority of species found in protected areas and, in some cases, changes brought in by productive activities may shift species composition towards higher biodiversity. In addition, production forests can be an asset for local populations and will therefore be respected.

These actions will be feasible only if they are integrated in a broader approach to solving the main conflicts affecting biodiversity conservation.

Protected Areas

Developing the system of protected areas means strengthening protection and management of existing areas and extending the system to new areas.

It is necessary to enhance the actual protection of existing protected areas with the objective of reaching effective total protection of TPAs and of bringing WRs under stricter regulations. Actions needed include marking boundaries, informing and educating local populations, strengthening and training the guard force and developing and implementing a management plan with the participation of local populations. Buffer zones should be created whenever appropriate. These buffer zones should be designed as areas of enhanced management of natural resources where sustained logging and/or hunting could be developed. Tourism should be promoted to generate revenues for local populations.

There is an urgent need to extend the system of protected areas in order to take into account critical sites that need protection. The above-mentioned IUCN regional programme for conservation and rational utilisation of forest ecosystems in Central Africa (IUCN 1980) has identified 61 critical sites for biodiversity conservation in Cameroon (21), Congo (12), Gabon (9), and Zaire (19). Critical sites are also being identified in Côte d'Ivoire to complement existing TPAs in the remaining coastal forests. Possibilities to develop protected areas in southwestern Ghana have also been recently identified. These sites should be progressively brought under management as TPAs or WRs within the coming decade and subsequently protected and managed under the conditions envisioned for the existing protected areas. A specific review of partially protected areas set aside for use by forest dwelling people should also be undertaken to identify possible additional areas to protect.

Permanent Multi-Purpose Forests

Developing a multi-purpose permanent forest estate means converting a large portion of the "Domaine Protégé" into gazetted forests, which will be managed for production and protection of the environment (soils, water, and microclimate) with due consideration to biodiversity conservation. This is an urgent but also immense task. Indeed, this means gazetting, protecting and progressively managing, within the coming decade, more than ten times the area of forests presently gazetted for production and protection. It should be noted that the latter were established over more than four decades. This process has started in Ghana and Côte d'Ivoire. The bulk of the task pertains to Cameroon, Congo, Gabon, and Zaire. It will include the following activities: designing the system, gazetting and demarcating the forests, surveying the potential of the forest, developing a management plan, and implementing and monitoring the management plan. All these activities need to be carried out with local participation.

A system of possible multi-purpose forests should be based on a land-use planning exercise which should be an expression of national vision and consensus regarding what land should be totally protected, what should be devoted to permanent productive and protective forests and what should be converted to agriculture. This system should complement and buffer the networks of TPAs.

Demarcating individual forests should be done after detailed studies of the socio-economic environment of each identified forest. Gazetting the forests as state or community forests should be the result of a thorough dialogue between local populations, traditional authorities, government services, and NGOs. Boundaries should be subsequently marked.

Surveying the potential of the forest should take into account timber and non-timber products, wildlife and other biological values. Forest inventory methods are well known and used but new, practical and cost-effective methods for comprehensive surveys of forest resources need to be disseminated.

The management plan of the forest should be designed for an initial period of 20 years, but should be subject to successive revisions. The forest will be divided into Working Circles (WCs) which are designed to meet specific objectives. The WCs which are the most likely to be determined are the following:

1) **Conservation WCs** to preserve biodiversity in critical sites, were regulations will be similar to those in effect in TPAs.

2) **Protection WCs** to protect fragile environments, which will be totally protected.

3) **Production WCs** which will have as primary objectives the production of timber and non-timber products on a sustainable basis (regulations for timber production are suggested in Table 2 and are based on the proposed ITTO guidelines for biodiversity conservation in production forests. Regulations for non-timber products will aim at better organised harvesting by local people).

4) **Rehabilitation WCs** where the objective will be to restore the forest cover in degraded areas by natural regeneration where possible, otherwise by plantations.

5) **Agroforestry WCs** where agriculture activities will be permitted but will be limited to stable combinations of annual and perennial crops.

Implementation of the management plan will be based on contracts between government services, local communities, NGOs and the private sector. Monitoring the implementation of the management plan will pay particular attention to the impact of logging on the ecological functions of the forest.

Conflicts and Proposed Solutions

The main issues regarding biodiversity conservation are closely related to the conflicts between conservation and other uses of TMF resources.

The first and main conflict is forest conservation and management versus land for farmers. This conflict can be solved in favour of conservation only if farmers intensify their production systems and if they benefit from forest conservation and management. Farmers must be granted security of land tenure to encourage investment in sustainable agriculture. The government should provide incentives to encourage intensification of agriculture. Local farmers should benefit financially from forest conservation by earning money from the tourist trade or by being granted exploitation rights for timber or non-timber products.

It should be noted, however, that economically attractive and environmentally feasible agricultural technologies to increase production per unit area are not generally available to the people who farm in TMF areas. Much research needs to be done to develop improved farm systems in these zones.

The second conflict is forest management for production versus forest conservation. This conflict can be solved in favour of conservation if productive uses of TMF are put under sustained management regulations and if operators (mainly private companies) have an interest in sustained management of forest resources. To resolve this conflict, the above-mentioned comprehensive regulations for multi-purpose management of TMFs should be implemented and the capacity to enforce these regulations should be strengthened. Further benefits would accrue by developing incentives ("green" labelling, promotion of lesser known species) for better marketing of products from forests managed on a sustained basis.

To solve these two conflicts, governments have the central responsibility of developing policy reforms and incentive systems. They will have to share the responsibility of implementing conservation and management programmes with local populations, NGOs, producers groups and the private sector. Commitment of governments to do this will be based on a clear understanding of the broad economic values of the role of TMF in sheltering biodiversity, protecting soils and waters, and producing goods.

5. Proposed Strategy

The strategy to achieve biodiversity conservation in TMF should combine the following:

1)   Developing a national land-use planning policy in areas covered by TMF.

2)   Developing policy reforms.

3)   Carrying out training and research activities.

4)   Starting a programme of progressive field operations.

Land-use Planning

A land-use planning policy should include a national vision of the best use of the resources of TMF areas that will guide sectoral policies, and operational mechanisms to put this vision into effect.

The land-use planning policy should begin with developing an agro-ecological zoning for the TMF areas in each country, which will serve as a framework for land-use planning and natural resource management. This agro-ecological zoning should determine areas that need to be fully protected for biodiversity conservation and zones that should be preserved for forest dwellers. This zoning should also identify areas that should stay under permanent forest cover and be managed for production and/or protection and areas best suited for sustainable agricultural development, including agroforestry and livestock. This agro-ecological zoning needs to be consistent with other government sectoral policies (e.g. population, urbanisation, and transportation infrastructure). The zoning exercise should adopt a multi-disciplinary approach and will be developed through mechanisms that ensure inter-ministerial consensus.

The agro-ecological zoning should serve as a framework for sectoral policies. As regards biodiversity conservation and related forestry policies, a national system of protected areas and multi-purpose gazetted forests should be designed according to the zoning and a programme to establish this system should be proposed. Regulations regarding the uses of TMFs remaining in the zones identified for agriculture should be established. Policies to develop a system of industrial wood processing units should be devised in accordance with the above zoning.

The key to successful land-use planning policy will be to develop instruments to render it effective. These instruments include methods and mechanisms for participative planning of land use and natural resource utilisation in TMF areas. Other instruments to be developed are:

1)   the institutional and regulatory framework to implement land-use plans;

2)   a system to monitor land use and natural resource utilisation in the TMF area; and

3)   the challenge of increasing returns to agriculture labour in marginal environments.

Policy Reforms

Policy reforms are crucial to improve the legal, fiscal, and institutional environment of any future effort to preserve and manage TMFs. Fiscal reforms and new incentives are essential to finance this effort.

Existing legislation should be amended in order to:

1)   achieve a coherent set of legal provisions and regulations regarding land tenure and use of natural forest resources;

2)   allow local populations to participate in conservation and management of TMFs. Private companies should be involved in management of production forests;

3)   clarify the concept and role of buffer zones around protected areas; and

4)   be more specific about the content of management plans and about technical regulations regarding forest exploitation.

Existing systems of forest revenues and forest taxes should also be amended in order to:

1)   increase present forest revenues in line with the actual value of timber resources;

2)   improve the actual collection of taxes and channel back financial resources into forest conservation and management; and

3)   make local populations benefit from a share of forest revenues.

Recent studies have concentrated on points (1) and (2) in the countries under review. Point (3) needs to be addressed also.

Incentives to preserve TMFs and to manage production forests on a sustained basis should be developed. Realistic mechanisms by which the international donor community would compensate governments for setting aside potentially productive zones of TMF should be thoroughly explored in the socio-economic context of the countries under review. The same applies to mechanisms by which governments would compensate local populations for not encroaching on protected areas. Preferential access to international markets for timber from managed permanent forests should also be promoted through "green" labelling.

Government institutions should be reorganised in accordance with the roles and responsibilities that should be given to local communities, NGOs and private companies in TMF conservation and management. Mechanisms to better integrate these institutions with other institutions responsible for rural development and to better coordinate these institutions with those in charge of energy, transportation and industry should be developed. Financial resources should be reallocated to strengthen reorganised and decentralised government institutions in implementing their respective tasks which include:

1)   planning and monitoring activities;

2)   providing high level technical assistance to other partners in TMF conservation and management; and

3)   enforcing regulations.

Training, Research, and Studies

A tremendous effort is needed to train and educate local populations, government agents, private entrepreneurs, and decision makers in matters related to biodiversity conservation in TMFs. This effort pertains to technical, social, economic, and cultural subjects and needs to be carefully designed. An immediate step would be to include biodiversity conservation and sustained management of forest resources in ongoing curricula with special emphasis on forestry field staff training.

An equal effort is needed to strengthen research in order to better understand socio-cultural factors regarding TMF conservation and management, to devise methods to evaluate the economic benefits from forest conservation and management and to increase technical knowledge on how TMF ecosystems function and how they should be managed. This should be achieved by strengthening national research institutions, developing pilot operations and promoting networking between national institutions to other relevant international research organisations.

Country reviews have recommended the following studies:

1)   Study of the impact of logging on biodiversity in general and, in particular, on natural forest regeneration in TMF of Western and Central Africa.

2)   Systematic country surveys of biodiversity and endangered species.

3)   Assessments of the status of present knowledge on economic values of the role of TMFs in sheltering biodiversity, protecting soils and waters, and producing goods.

Field Operations

Field operations should be developed in two successive phases. A first phase would start developing urgent actions to protect critical sites and establish pilot operations to manage multi-purpose gazetted forests. In the meantime, the associated land-use planning, policy reforms and studies would be developed. A second phase would develop the systems of extensive protected areas and permanent forests managed for multiple purposes on the basis of the results of the land-use planning exercise. These activities would take place in an improved legal, fiscal and institutional environment.

Under phase one, existing TPAs and WRs would be managed and more efficiently protected. Management plans would be prepared for new areas around critical sites already identified. Multi-purpose management of pilot gazetted forests would be developed in the broader context of pilot land-use planning operations. Under these pilot operations, management methods would be tested and new mechanisms would be designed to stimulate local participation. During this phase, logging concessions should be issued according to a careful and conservative programme in order to limit possible future incompatibilities with the land-use planning exercises. The cost of these actions will be low and could be met by external financing with existing financial instruments.

Under phase two, larger programmes to progressively protect and manage a comprehensive system of TPAs and WRs would be developed in the zones identified for total protection by the land-use planning exercise. Similar large programmes to establish and manage systems of multi-purpose permanent forests would be progressively developed in zones meant to stay under permanent forest cover.

While waiting for the gazettement of these zones, logging concessions should be issued under a system similar to the one developed in Congo with the "Unités Forestières d'Aménagement". The cost of these programmes will be much higher and could only be met by effective participation of local populations, increased fiscal revenues from productive forest activities and new international compensation mechanisms. All this is dependent on the results of phase one.

6. Possible Role of ITTO

The possible role of ITTO to influence policies and programmes relating to timber production and biodiversity conservation should be to stimulate an international debate on relevant issues, support the design and adoption of policy reforms and incentives, conduct studies, coordinate pilot operations and provide technical assistance.

Advisory Group: ITTO should be instrumental in establishing an independent advisory group to promote dialogue between governments, conservation organisations and the private sector involved in industrial forestry. This group should stimulate the international debate on the conservation implications of the timber trade.

Support to Fiscal Reforms: ITTO should help African governments design fiscal reforms with a focus on realistic mechanisms to make local populations benefit from a share of forest revenues. ITTO could sponsor a regional study that would end up with practical recommendations to establish these mechanisms.

Incentives: ITTO should help determine whether compensating governments for setting aside potentially productive zones of TMF is desirable and feasible in the economic context of African countries. ITTO should also promote preferential access to international markets for African timber from managed permanent forests through "green" labelling.

Studies: ITTO should conduct a regional study on the impact of logging on biodiversity in general and, in particular, on natural forest regeneration in TMFs of Western and Central Africa. This study should help develop guidelines for sustained management of TMF. ITTO should assist African countries in developing systematic country surveys of biodiversity and endangered species. ITTO should also make an assessment of present knowledge regarding economic values of the role of TMF in sheltering biodiversity, protecting soils and waters, and producing goods.

Network of managed multi-purpose forests: ITTO should help establish and coordinate a network of multi-purpose gazetted forests managed for production and protection in TMF of Western and Central Africa. The initial objective of this network is to share experience and transfer methods and techniques from one country to an other.

Surveys of TMF Resources: ITTO should provide technical assistance in developing new, practical and cost-effective methods for comprehensive surveys of forest resources. These surveys should be designed to be an operational basis for preparation of management plans for TPAs or multi-purpose permanent forests.

Guidelines to Develop Management Plans: ITTO should design and promote practical guidelines for developing management plans for multi-purpose gazetted forests in African TMF. These guidelines should be presented as a manual which would integrate, inter alia, the ITTO guidelines on biodiversity conservation in production forests.

TABLE 1

(Million of ha)	Cameroon	Congo	C.I.	Gabon	Ghana	Zaire
Country Total Area	47.5	34.2	32.2	268	23.9	234.5
TMF Total Area	20.0	22.4	2.5	23.0	2.1	105.8
Totally Protected Areas (TPA)	0.7	0.3	0.5	–	–	6.0
Wildlife Reserves	0.4	0.4	0.1	1.8	0.1	1.7
Gazetted Forests other than TPA and WR	1.3	0.0	1.4	–	1.8	0.5
"Domaine Protege"	17.6	21.7	0.5	21.2	0.2	97.6
Deforestation/year (1000 ha)	80	22	290	15	22	180

Remarks:
1. Data regarding Togo have not been included here because TMF are marginal.
2. The area of TPA in TMF in Ghana and Gabon are respectively 18,000ha and 10,000ha.

TABLE 2 PROPOSED REGULATIONS FOR PRODUCTION WORKING CIRCLES IN MANAGED FORESTS

The following regulations are proposed regarding timber production in the Production Working Circles that will be determined by management plans:

Felling Cycle: Polycyclic felling will be adopted; each cycle will range from 20 to 40 years according to the potential of the forest (e.g. closer to 20 years in areas rich in rather fast growing species such as *Triplochiton scleroxylon, Aucoumea klaineana,* and closer to 40 years in areas rich in *Meliaceae*).

Time-zoning of logging: Exploitation of timber will be authorised during one fourth or one fifth of the cycle on a "felling unit" covering one fourth or one fifth of the Production WC, to allow the forest (plant and wildlife) to recuperate.

Annual Cut: The annual allowed cut will be determined in accordance with the estimated potential and annual yield of the "felling unit", on the basis of preliminary management inventories.

Roads: Roads inside the forest will be opened and closed according to time-zoning of logging.

Total Protection: Sites of specific importance for biodiversity conservation will be totally protected in the Production WC. Species of particular biological and ecological importance will also be totally protected.

Best Forestry Practice Regulations: Best practice regulations will include a comprehensive pre-exploitation survey, mapping, standards of road planning and construction, standards of planning of log extraction routes, standards about maximum wood waste in felled trees, regulations to limit felling damage (directional felling, climber cutting), an audit of regeneration before felling, and a post-felling assessment of stands.

Silvicultural Treatments: Silvicultural treatments will be only tested, during the first 10 years of the management plan, in an "Experimental Sub-Working Circle" established within the Production WC.

Bibliography

Amine, M. and Besong, J. 1990. Untitled. Unpublished report prepared for the IUCN Forest Conservation Programme workshop, "Realistic Strategies for Tropical Forest Conservation" in Perth, Australia.

FAO. 1988. *An Interim Report on the State of Forest Utilisation in the Developing Countries.* FO:MISC/88/7. FAO, Rome, Italy. 18pp.

IUCN. 1986. *Review of the Protected Areas System in the Afrotropical Realm.* IUCN, Gland, Switzerland and Cambridge, UK. 256pp.

IUCN. 1989. *La Conservation des Ecosystèmes forestiers d'Afrique Centrale.* IUCN, Gland, Switzerland and Cambridge, UK.

IUCN. 1990a. *1990 United Nations List of National Parks and Protected Areas.* IUCN, Gland, Switzerland and Cambridge, UK.

IUCN. 1990b. Untitled. Unpublished report prepared by IUCN staff for the IUCN Forest Conservation Programme workshop, "Realistic Strategies for Tropical Forest Conservation" in Perth, Australia.

Kasali, L. 1990. Untitled. Unpublished report prepared for the IUCN Forest Conservation Programme workshop, "Realistic Strategies for Tropical Forest Conservation" in Perth, Australia.

Memvié, J.B. 1990. Untitled. Unpublished report prepared for the IUCN Forest Conservation Programme workshop, "Realistic Strategies for Tropical Forest Conservation" in Perth, Australia.

Nadjombe, O. 1990. Untitled. Unpublished report prepared for the IUCN Forest Conservation Programme workshop, "Realistic Strategies for Tropical Forest Conservation" in Perth, Australia.

N'Sosso, D. 1990. Untitled. Unpublished report prepared for the IUCN Forest Conservation Programme workshop, "Realistic Strategies for Tropical Forest Conservation" in Perth, Australia.

Rietbergen, S. 1988. Natural forest management for sustainable timber production: the Africa region. Unpublished report prepared for IIED and ITTO.

Sayer, J.A., Harcourt, C., and Collins, M.N. (in press). *The Conservation Atlas of Tropical Forests: Africa.* Macmillan Press Ltd., London.

Tufour, K. 1990. Untitled. Unpublished report prepared for the IUCN Forest Conservation Programme workshop, "Realistic Strategies for Tropical Forest Conservation" in Perth, Australia.

Wilks, C. 1990. *La Conservation des Ecosystèmes forestiers du Gabon.* IUCN, Gland, Switzerland and Cambridge, UK. 215pp.

World Resources Institute (WRI). 1990. *World Resources: 1990-91.* Oxford University Press. Oxford. 383pp.

ANNEX

ASIA

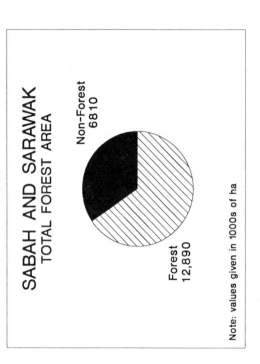

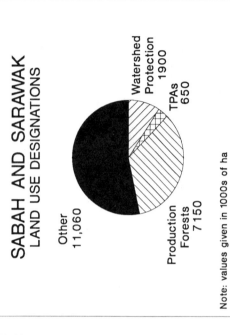

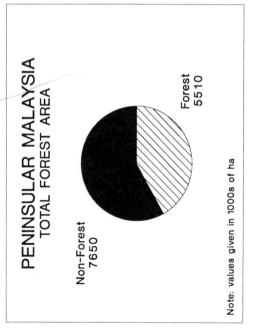

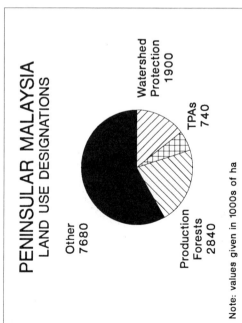

ASIA

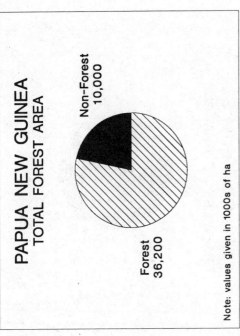

NEPAL
TOTAL FOREST AREA

Non-Forest
8162

Forest
5518

Note: values given in 1000s of ha

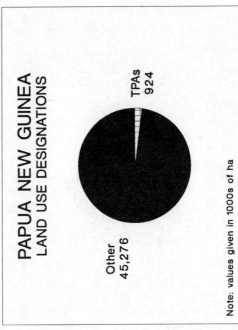

PAPUA NEW GUINEA
TOTAL FOREST AREA

Non-Forest
10,000

Forest
36,200

Note: values given in 1000s of ha

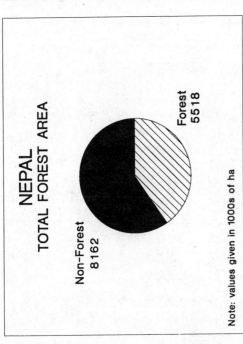

NEPAL
LAND USE DESIGNATIONS

TPAs
958.5

Other
12,721.5

Note: values given in 1000s of ha

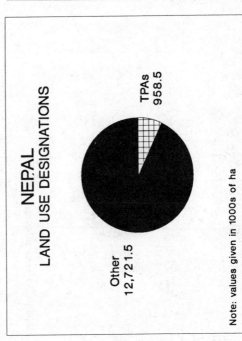

PAPUA NEW GUINEA
LAND USE DESIGNATIONS

TPAs
924

Other
45,276

Note: values given in 1000s of ha

235

ASIA

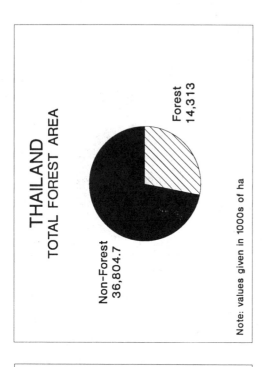

THAILAND
TOTAL FOREST AREA

Forest
14,313

Non-Forest
36,804.7

Note: values given in 1000s of ha

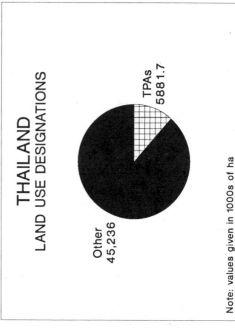

THAILAND
LAND USE DESIGNATIONS

TPAs
5881.7

Other
45,236

Note: values given in 1000s of ha

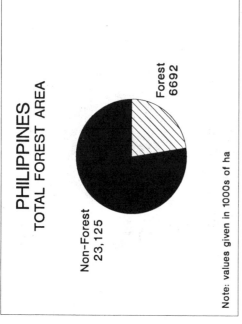

PHILIPPINES
TOTAL FOREST AREA

Forest
6692

Non-Forest
23,125

Note: values given in 1000s of ha

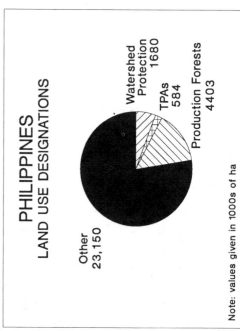

PHILIPPINES
LAND USE DESIGNATIONS

Watershed
Protection
1680

TPAs
584

Production Forests
4403

Other
23,150

Note: values given in 1000s of ha

LATIN AMERICA

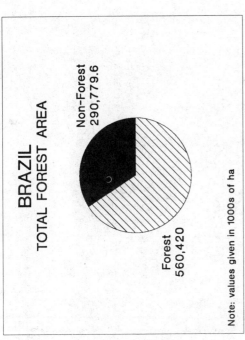

BRAZIL
TOTAL FOREST AREA

Non-Forest
290,779.6

Forest
560,420

Note: values given in 1000s of ha

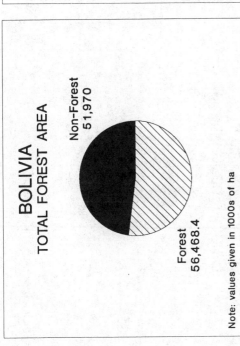

BOLIVIA
TOTAL FOREST AREA

Non-Forest
51,970

Forest
56,468.4

Note: values given in 1000s of ha

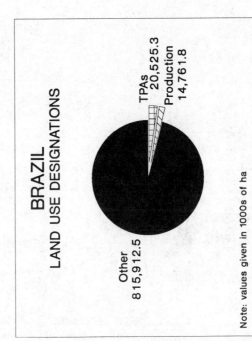

BRAZIL
LAND USE DESIGNATIONS

TPAs
20,525.3

Production
14,761.8

Other
815,912.5

Note: values given in 1000s of ha

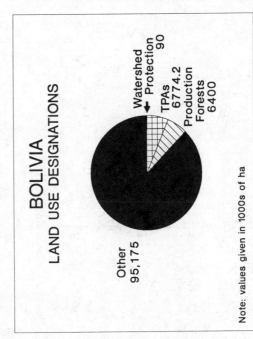

BOLIVIA
LAND USE DESIGNATIONS

Watershed
Protection 90

TPAs
6774.2

Production
Forests
6400

Other
95,175

Note: values given in 1000s of ha

LATIN AMERICA

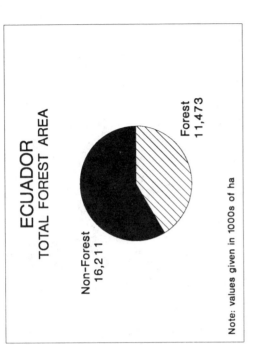

ECUADOR
TOTAL FOREST AREA

Forest
11,473

Non-Forest
16,211

Note: values given in 1000s of ha

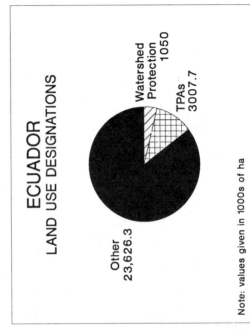

ECUADOR
LAND USE DESIGNATIONS

Watershed
Protection
1050

TPAs
3007.7

Other
23,626.3

Note: values given in 1000s of ha

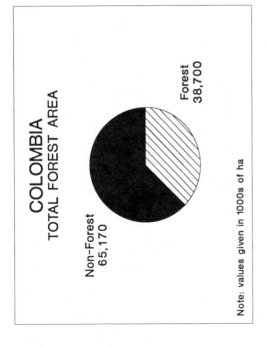

COLOMBIA
TOTAL FOREST AREA

Forest
38,700

Non-Forest
65,170

Note: values given in 1000s of ha

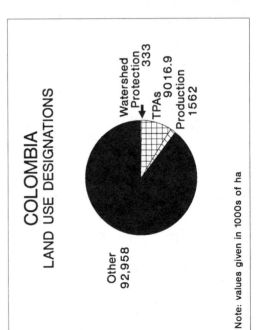

COLOMBIA
LAND USE DESIGNATIONS

Watershed
Protection 333

TPAs
9016.9

Production
1562

Other
92,958

Note: values given in 1000s of ha

LATIN AMERICA

HONDURAS
TOTAL FOREST AREA

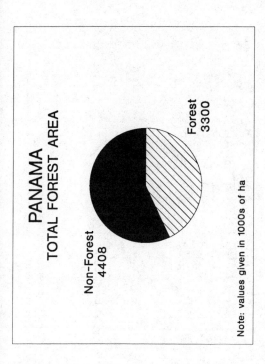

Non-Forest
6138

Forest
5051

Note: values given in 1000s of ha

PANAMA
TOTAL FOREST AREA

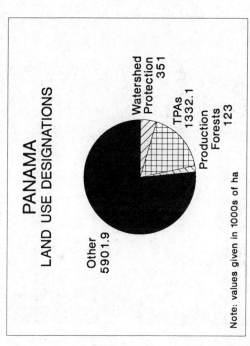

Non-Forest
4408

Forest
3300

Note: values given in 1000s of ha

HONDURAS
LAND USE DESIGNATIONS

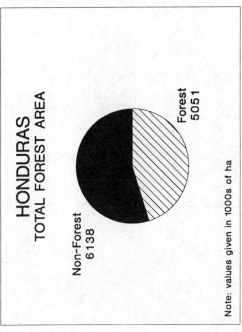

Other
8843.7

TPAs
703.3

Production
Forests
1642

Note: values given in 1000s of ha

PANAMA
LAND USE DESIGNATIONS

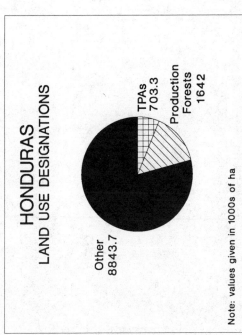

Other
5901.9

Watershed
Protection
351

TPAs
1332.1

Production
Forests
123

Note: values given in 1000s of ha

LATIN AMERICA

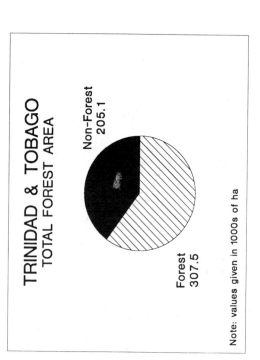

TRINIDAD & TOBAGO
TOTAL FOREST AREA

Non-Forest
205.1

Forest
307.5

Note: values given in 1000s of ha

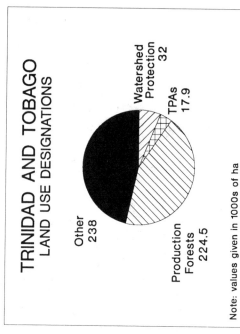

TRINIDAD AND TOBAGO
LAND USE DESIGNATIONS

Watershed
Protection
32

TPAs
17.9

Other
238

Production
Forests
224.5

Note: values given in 1000s of ha

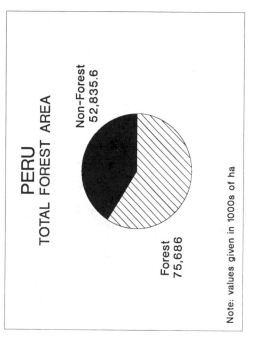

PERU
TOTAL FOREST AREA

Non-Forest
52,835.6

Forest
75,686

Note: values given in 1000s of ha

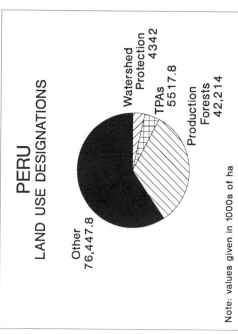

PERU
LAND USE DESIGNATIONS

Watershed
Protection
4342

TPAs
5517.8

Production
Forests
42,214

Other
76,447.8

Note: values given in 1000s of ha

AFRICA

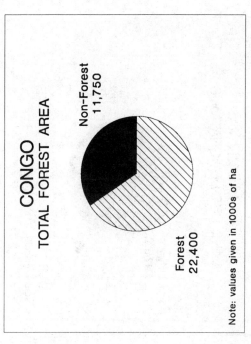

CAMEROON
TOTAL FOREST AREA

Non-Forest
29,040

Forest
17,500

Note: values given in 1000s of ha

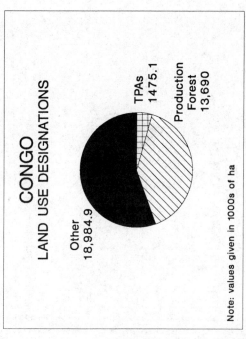

CONGO
TOTAL FOREST AREA

Non-Forest
11,750

Forest
22,400

Note: values given in 1000s of ha

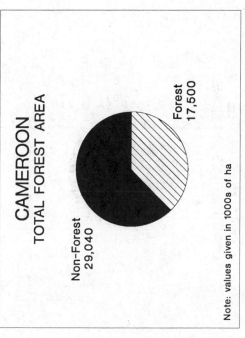

CAMEROON
LAND USE DESIGNATIONS

Watershed
Protection 69
TPAs 2099.7
Production
Forests
1262.1

Other
43,109.2

Note: values given in 1000s of ha

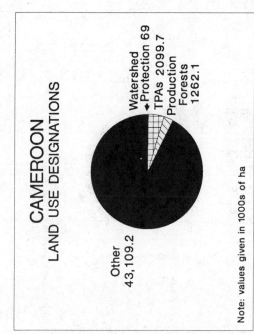

CONGO
LAND USE DESIGNATIONS

TPAs
1475.1
Production
Forest
13,690

Other
18,984.9

Note: values given in 1000s of ha

AFRICA

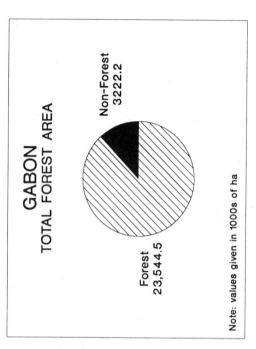

GABON
TOTAL FOREST AREA

Non-Forest
3222.2

Forest
23,544.5

Note: values given in 1000s of ha

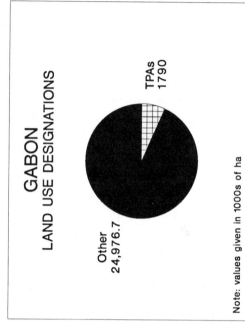

GABON
LAND USE DESIGNATIONS

TPAs
1790

Other
24,976.7

Note: values given in 1000s of ha

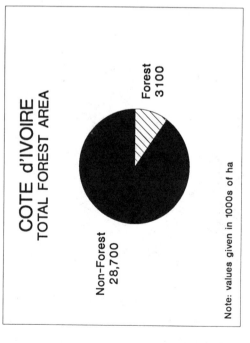

COTE d'IVOIRE
TOTAL FOREST AREA

Forest
3100

Non-Forest
28,700

Note: values given in 1000s of ha

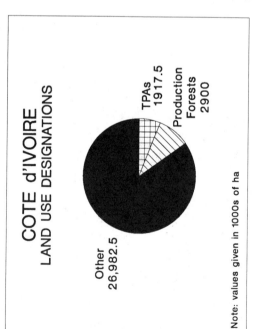

COTE d'IVOIRE
LAND USE DESIGNATIONS

TPAs
1917.5

Production
Forests
2900

Other
26,982.5

Note: values given in 1000s of ha

AFRICA

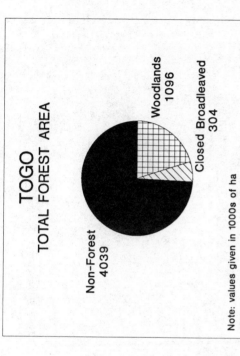

TOGO
TOTAL FOREST AREA

Woodlands 1096

Closed Broadleaved 304

Non-Forest 4039

Note: values given in 1000s of ha

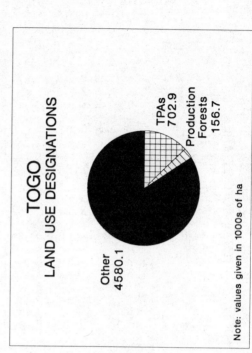

TOGO
LAND USE DESIGNATIONS

TPAs 702.9

Production Forests 156.7

Other 4580.1

Note: values given in 1000s of ha

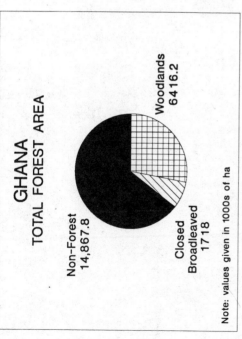

GHANA
TOTAL FOREST AREA

Woodlands 6416.2

Non-Forest 14,867.8

Closed Broadleaved 1718

Note:: values given in 1000s of ha

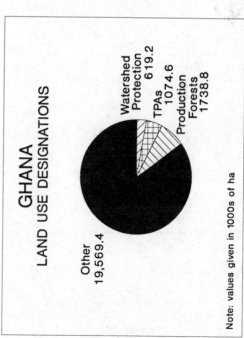

GHANA
LAND USE DESIGNATIONS

Watershed Protection 619.2

TPAs 1074.6

Production Forests 1738.8

Other 19,569.4

Note: values given in 1000s of ha

AFRICA

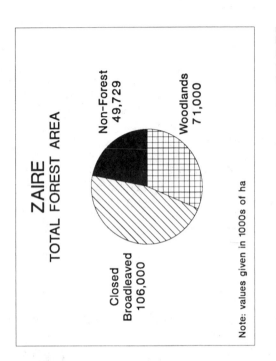

ZAIRE
TOTAL FOREST AREA

Non-Forest
49,729

Woodlands
71,000

Closed
Broadleaved
106,000

Note: values given in 1000s of ha

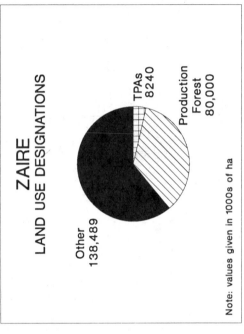

ZAIRE
LAND USE DESIGNATIONS

TPAs
8240

Production
Forest
80,000

Other
138,489

Note: values given in 1000s of ha